Animal Nutrition and Welfare in Sustainable Production Systems

Animal Nutrition and Welfare in Sustainable Production Systems

Editors

Nikola Puvača
Vincenzo Tufarelli
Eva Voslarova

MDPI • Basel • Beijing • Wuhan • Barcelona • Belgrade • Manchester • Tokyo • Cluj • Tianjin

Editors

Nikola Puvača
Department of Engineering
Management in Biotechnology
University BA in Novi Sad
Novi Sad
Serbia

Vincenzo Tufarelli
Department of DETO
University of Bari Aldo Moro
Valenzano
Italy

Eva Voslarova
Department of Animal
Protection and Welfare and
Veterinary Public Health
University of Veterinary Sciences
Brno
Czech Republic

Editorial Office
MDPI
St. Alban-Anlage 66
4052 Basel, Switzerland

This is a reprint of articles from the Special Issue published online in the open access journal *Sustainability* (ISSN 2071-1050) (available at: www.mdpi.com/journal/sustainability/special_issues/Animal_nutririon_welfare).

For citation purposes, cite each article independently as indicated on the article page online and as indicated below:

LastName, A.A.; LastName, B.B.; LastName, C.C. Article Title. *Journal Name* **Year**, *Volume Number*, Page Range.

ISBN 978-3-0365-2819-9 (Hbk)
ISBN 978-3-0365-2818-2 (PDF)

Contents

About the Editors

Nikola Puvača

Nikola Puvača is a Researcher and Associate Professor at the Department of Engineering Management in Biotechnology of the University Business Academy in Novi Sad, Serbia. He is a multidisciplinary scientist with Ph.D. diplomas in Biotechnology, Veterinary Medicine, and Human Medicine, and a postdoctoral diploma in Toxicology and Molecular Genetics. He has significant experience in animal and poultry science, with a major interest in nutrition, feed quality and safety, natural alternatives for antibiotics, and antimicrobial resistance. He is involved in many research collaborations in various scientific fields. He currently serves as an Editorial Board Member and peer reviewer for many indexed journals, and he is the author of more than 200 scientific papers published in international journals and the proceedings of national and international conferences. He is a guest lecturer in several international universities and has won many rewards for scientific and professional work. Currently, he holds the position of Vice Dean for Internarial Relations at his home faculty.

Vincenzo Tufarelli

Vincenzo Tufarelli is an Associate Professor of Animal Nutrition at the Department of Emergency and Organ Transplants (DETO), Section of Veterinary Science and Animal Production of the University of Bari "Aldo Moro", Italy. He has considerable experience in animal and poultry science, with a particular interest in nutrition and feed technology. He is involved in many research collaborations, even with international institutions, in the field of animal science and feed quality. He serves as an editorial board member and peer reviewer for many indexed journals and he is the author of more than 200 scientific papers published in international journals and proceedings of national and international conferences.

Eva Voslarova

Prof. Ing. Eva Voslarova, Ph.D. graduated from the Mendel University Brno (Ing.) in 2001. She obtained a Ph.D. degree in Veterinary Public Health and Animal Protection (2004), habilitation (Doc.) in Animal Protection and Welfare (2008) and full professorship (Prof.) in Animal Protection and Welfare (2019) at the University of Veterinary Sciences Brno. As a university professor and the Head of Animal Protection and Welfare Unit at the University of Veterinary Sciences Brno, she is responsible for animal protection and welfare courses for both pre-graduate and post-graduate students and other welfare courses within lifelong education, including the further education of veterinary inspectors. She has international teaching experience as a visiting professor at several European universities. She was also repeatedly invited as a tutor for the EU courses Better Training for Safer Food. The results of her research aiming to ensure the welfare of farm, companion and captive-kept wild animals have been published in peer-reviewed scientific journals worldwide (751 SCI citations; h-index = 18). She has been involved in numerous national research projects focused on the welfare of animals and/or food safety and quality. From 2015 to 2017, she was the work package leader for the Horizon 2020 Hennovation project. She is the National representative for cooperation with EFSA in the area of animal welfare, dairy, and a member of the EFSA working group on transport of caged species. Since 2005, she has been a coordinator of the International Animal Protection and Welfare Conference hosted annually by the University of Veterinary Sciences in Brno, Czech Republic.

Preface to "Animal Nutrition and Welfare in Sustainable Production Systems"

Today, food animal production systems demand high energy, land, chemicals, and water—all of which are increasingly becoming scarce. Thus, change and innovation are required in many animal production systems to meet the present and future demands for animal products sustainably. Over the last four decades, inexpensive grain, energy, and protein have enabled the economic development of intensive meat, eggs, and milk production systems based on feeding grains and other ingredients sourced from far-off places. The poultry and pig intensive production systems have become highly capital intensive, and they have resulted in many environmental challenges.

If the price of feedstuffs rises above a critical level, they might even become economically and environmentally unviable. These issues may become magnified due to increasing competition for arable land for food, feed, and biofuel production. So far, in many situations, the feed has been produced, and nutrition has been balanced to achieve maximum production, with high economic benefits. However, it has also contributed to ecosystem degradation and global warming through methane and nitrous oxide. Moreover, animal production results in increased energy consumption at every step, and even more so in intensive systems. These situations demand attention to examine the excessive use of resources and consider ways to adopt more efficient processes and procedures.

Food animal feed, nutrition, and welfare are the foundation of successful animal systems. They directly or indirectly affect the entire animal production sector, associated services, public goods, and services, including animal productivity, health and welfare, product quality and safety, land use and land-use change, and greenhouse gas emissions. The sustainability of food animal nutrition and welfare is crucial in developing animal production across production systems. The sustainable increase in animal productivity, which is key to meeting the large current and future demands for animal origin products, cannot be achieved without sustainable animal nutrition and welfare.

Sustainable animal nutrition and welfare are expected to be beneficial for the food animals, the environment, and society. Additionally, they are likely to generate socioeconomic benefits, furthering poverty alleviation and food security efforts. This requires researchers, extension workers, science managers, policy makers, industry, and farmers.

Nikola Puvača, Vincenzo Tufarelli, Eva Voslarova
Editors

sustainability

Article

Effects of Using Farm-Grown Forage as a Component in ad Libitum Liquid Feeding for Pregnant Sows in Group-Housing on Body Condition Development and Performance

Clara Berenike Hartung [1,*], Stephanie Frenking [1], Bussarakam Chuppava [1], Friederike von und zur Mühlen [1], Josef Kamphues [1], Peter Ebertz [2], Richard Hölscher [3], Eva Angermann [4] and Christian Visscher [1]

[1] Institute for Animal Nutrition, University of Veterinary Medicine Hannover, Foundation, Bischofsholer Damm 15, D-30173 Hanover, Germany; sekretariat-tierernaehrung@tiho-hannover.de (S.F.); Bussarakam.chuppava@tiho-hannover.de (B.C.); service-tierernaehrung@tiho-hannover.de (F.v.u.z.M.); josef.kamphues.iR@tiho-hannover.de (J.K.); Christian.Visscher@tiho-hannover.de (C.V.)
[2] Institute of Agricultural Engineering, University of Bonn, Nussallee 5, D-53115 Bonn, Germany; ebertz@uni-bonn.de
[3] Hoelscher + Leuschner GmbH & Co. KG, D-48488 Emsbueren, Germany; R.Hoelscher@hl-agrar.de
[4] Institute of Agricultural and Nutritional Sciences, Department of Animal Husbandry and Ecology, Martin-Luther-University Halle-Wittenberg, Theodor-Lieser-Str. 11, D-06120 Halle, Germany; Eva-Angermann@gmx.de
* Correspondence: clara.berenike.hartung@tiho-hannover.de

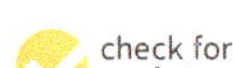
check for **updates**

Citation: Hartung, C.B.; Frenking, S.; Chuppava, B.; von und zur Mühlen, F.; Kamphues, J.; Ebertz, P.; Hölscher, R.; Angermann, E.; Visscher, C. Effects of Using Farm-Grown Forage as a Component in ad Libitum Liquid Feeding for Pregnant Sows in Group-Housing on Body Condition Development and Performance. *Sustainability* **2021**, *13*, 13506. https://doi.org/10.3390/su132413506

Academic Editors: Nikola Puvača, Vincenzo Tufarelli and Eva Voslarova

Received: 29 October 2021
Accepted: 4 December 2021
Published: 7 December 2021

Publisher's Note: MDPI stays neutral with regard to jurisdictional claims in published maps and institutional affiliations.

Abstract: When feeding pregnant sows, optimal body condition at birth is sought to avoid the effects of a deviant nutritional condition on health and performance. Various feeding concepts exist but mainly have a restriction in quantity and renunciation of farm-grown forage in common. An ad libitum liquid feeding system based on farm-grown forage in combination with a sow sorting gate (according to body weight—using mechanical scales) was realized on a commercial swine farm. The sorting gate coordinated access to two feeding areas with rations based on whole plant wheatsilage (WPWS) differing in energy content. In this study with a total of 183 pregnant sows, effects of restrictive dry feeding (System I) were compared with ad libitum liquid feeding based on farm-grown forage (System II). Sows were monitored regarding body condition development during pregnancy by measuring body condition score (BCS), body weight (BW), and back fat thickness (BFT) on different time points. Sow and piglet health (vaginal injuries of sows, rectal temperature during the peripartal period, vitality of newborn piglets) and performance data regarding litter characteristics were also recorded. Body condition development of the sows was absolutely comparable. Performance indicators and the course of birth were also similar but with significantly higher scores for piglet vitality in System II ($p < 0.05$). The tested concept offers opportunities for more animal welfare and sustainability but remains to be further investigated regarding the repertoire of possibly applied farm-grown forage and the effects of the concept in the transit phase of sows.

Keywords: forage; liquid feeding; gestation; lactation; sow; ad libitum feeding; sorting gate; feed self-sufficiency

1. Introduction

Developments regarding improvements in animal welfare in pig farming and feeding are being discussed throughout Europe [1,2]. Gestating sows must be kept in groups from the 29th day of pregnancy until 7 days before the determined farrowing time in accordance with the EU Council Directive 2008/120/EC of 18 December 2008.

In the feeding of pregnant sows, the aim is to achieve optimal body condition of the animals at birth to avoid the effects of an obese or very lean nutritional status on health and performance. This is because overconditioning can favor the occurrence of postpartum dysgalactia syndrome by prolonging the farrowing duration [3,4], and severe body weight (BW) losses can have negative effects on fertility and longevity [5,6].

To maintain optimal body condition in group-housed sows, only a limited number of housing concepts are suitable. For example, sows can be kept in small groups and fed restrictively according to condition. If the sows are kept in a large group, individual feeding can be used, for example, on automatic feeders [7]. To the best of our knowledge, there is as yet no management system of sows in large groups with ad libitum feed supply based on roughage and separation solely for feeding according to body condition.

The choice of feed components is varied in sow feeding. In the past—up to the second half of the last century—pigs were often fed roughage, silages, and beets, supplemented with concentrates and mineral feeds [8]; as pigs in far smaller production units were often housed outside and after the second world war, cereal-based feeds were very expensive [8]. In the intensified pig farming of the last few decades up to now, "combined feeding" has little significance because pigs are being expected to grow as efficiently and fast as possible [9]. Additionally to production performance, the labor and time requirements of the respective feeding system and problem-free manure removal are the basis for the commercial use of a husbandry system [10].

The current concentrate-based supply of feed meets the sows' nutritional needs, but not their desire for continuous foraging and voluntary feed intake [11–13]. Animals often show increased manipulation of housing equipment or stereotypic behavior due to an unfulfilled feeding and foraging motivation and hunger [14,15]. To ensure that the supply of crude fiber to pregnant gilts and sows is adequate, different crude fiber carriers and feeding concepts are available to pig farms [16–18]. Technologically advanced liquid feeding systems facilitate the safe provision of liquid diets from chopped whole plant silage in combination with other forages [19]. A basic forage-based and fiber-rich ration can meet nutrient requirements while as well meeting the sows' distinct food intake needs when an ad libitum diet is provided [13,20,21]. The gestating sows can consume an individual amount of feed together at any given time. Additionally, using higher amounts of farm-grown forage instead of imported concentrates can lead to higher feed self-sufficiency of farms [22,23]. Farm-grown forage used for ad libitum feeding of pregnant sows can therefore offer an option for animal nutrition and welfare in a sustainable production system.

The aim of the present study as part of the "SWOF" project (sow welfare optimized feeding) was to compare two feeding systems for group-housed sows in order to achieve ad libitum feeding in larger groups of sows while maintaining body condition during gestation, sow and piglet health in the peripartal period and performance data. The comparison included one conventional concentrate based and restricted dry feeding regime and one roughage-based ad libitum liquid feeding regime.

2. Materials and Methods

In a field trial, ad libitum feeding based on whole plant wheat-silage (WPWS) was implemented for pregnant sows in group housing using an automated sorting gate to give sows access to rations with either high or low energy content depending on their BW (System II). This was compared with a conventional feeding system (restrictively a dry complete pelleted feed, System I).

2.1. Animals

The animal housing and data collection took place in accordance with German regulations. No relevant interventions in accordance with the German Animal Welfare Act had been carried out on live animals (Approval by the Animal Welfare Officer of the University of Veterinary Medicine Hannover, Foundation, Hannover, reference: TVO-2017-V-49).

The study was carried out at a conventional farm in Germany with approximately 1200 sows (Danzucht) between January and June 2018. Sows were weaned at 1-week intervals. At a maximum of 28 days after service, sows were moved to the respective group-housing system. At day 108 of gestation, about one week before the estimated farrowing date (115 d p.c.), the sows entered the farrowing compartments. After a 28-day suckling period and weaning of the piglets, the sows again entered the service area. In two successive approaches, a total of 92 sows in System I and 91 in System II could be incorporated in the study. In the first run, 50 sows in System I and 49 in System II were included, in the second approach, 42 sows were included in each system. At the beginning of the trials, the sows have had an average of 2.86 litters in System I and 2.78 litters in System II.

2.2. Housing System

2.2.1. System I

In System I, each group consisting of 35–55 sows on the farm was housed in a compartment, which was again divided by a partition wall, where cross troughs were installed (Figure 1). Drinkers were mounted on the side walls in each compartment. Each pen in the compartments of the control group was equipped with volumetric feeders, and the cross trough extended the entire length of the compartment.

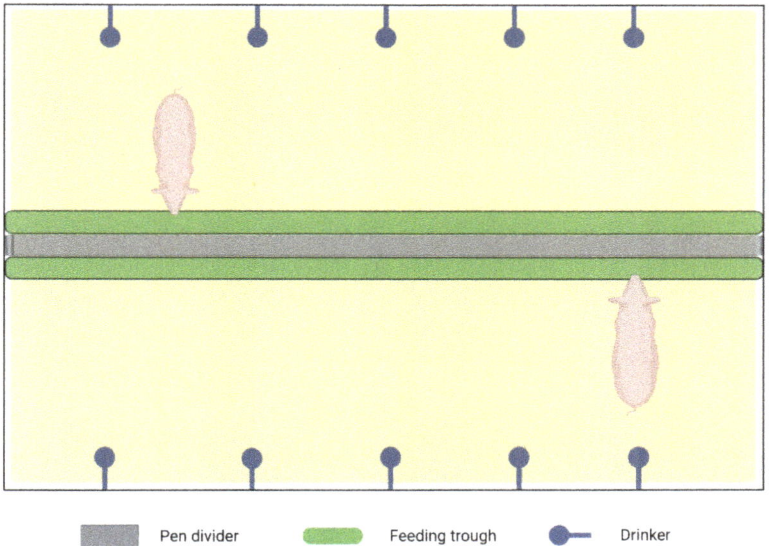

Figure 1. Housing of gestating sows in System I (Figure was created with BioRender.com) (accessed on 29 November 2021).

2.2.2. System II

In System II, the pregnant sows were kept in a large dynamic group, the size of which varied between 76 and 117 animals. Feasible group sizes in the two systems were determined by the structural conditions of the farm buildings. The sows were provided with different functional areas in the compartment—a resting/activity area and two feeding areas (Figure 2).

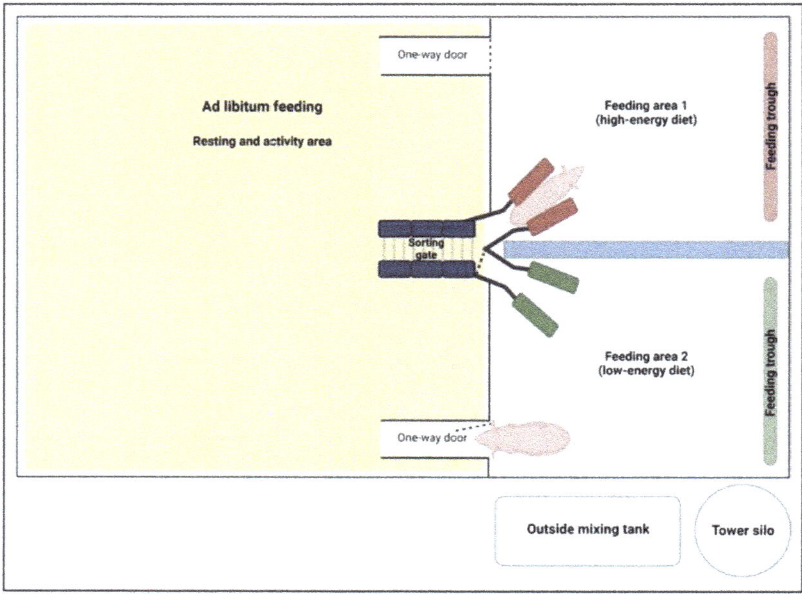

Figure 2. Housing of gestating sows in System II (Figure was created with BioRender.com) (accessed on 29 November 2021).

The resting/activity area was built with pen construction elements and had alternating perforated and paved areas. This functional area contained the drinkers and centrally the access to the feeding areas through a sorting gate (Hölscher + Leuschner GmbH & Co. KG®, Emsbüren, Germany). The sorting gate separated the feeding area from the activity and resting area (Figure 3).

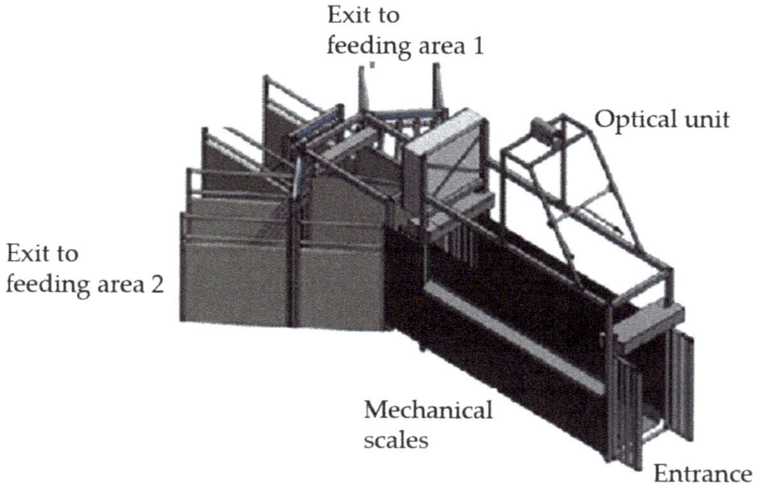

Figure 3. Design of the sorting gate for group-housed pregnant sows.

Transponder ear tags of the sows enabled automatic identification and storage of individual animal data. In addition to the optical unit (measuring height and width of the animals), the gate also contained mechanical scales for documenting BW. Sows could only enter the feeding areas through the gate, allowing access to one of two feeding areas with different feed ratios depending on body condition.

2.3. Diets and Feeding Technique

The sows in System I were fed restrictively using a chain feeding system with crumbled pellets (Tables 1 and 2). The amount of feed was manually adjusted on the volumetric feeders. Sows received feed twice daily. The complete feed used was purchased by the farm. In System II, the liquid feeding system from the company Hölscher + Leuschner GmbH & Co. KG (Emsbüren, Germany) was based on WPWS (Tables 1 and 2). The silage was taken from the silo and mixed with water in an outside mixing tank (approx. 11.5% DM). The mixture was pumped into the mixing tank of the liquid feeding system in the barn building and was completed with barley meal, soybean extraction meal, and mineral feed. Table 1 displays the botanical composition of the compound feeds used in Systems I and II. The two ad libitum rations were each pumped in stub lines to the troughs so that any ration could be fed at any time in System II. Sensors were integrated into each trough to measure the filling level. If the level in the trough was low, the feed was added automatically. Feeding was carried out in four feeding blocks and started in the morning at 03:00. A feeding break was taken from 21:00 to 03:00 to empty the trough once a day for hygienic reasons.

Table 1. Botanical composition of the complete feed for the restrictively fed sows in System I according to feed declaration and for the two ad libitum feed rations (% of dry matter-DM) in System II.

System I	System II		
Component	Component	Low-Energy Ration	High-Energy Ration
Barley	Barley	30.93	48.54
Wheat bran	WPWS	51.55	25.89
Oats	Soybean extraction meal	13.40	21.04
Rapeseed cake	Mineral feed	4.12	4.53
Extracted sunflower seed meal			
Sugar beet pulp			
Soybean hulls			
Soybeans (toasted)			
Malt culms			
Sugarbeet molasses			
Brewer's yeast			
Calcium carbonate			
Monocalcium phosphate			
Sodium chloride			

Table 2. Chemical composition of the three different compound feeds used in the trials (% of dry matter—DM).

Chemical Component	Restrictive Dry Feed-in System I	Ad Libitum Liquid Feed in System II	
		Low-Energy Ration	High-Energy Ration
Crude protein	17.2	18.5	22.5
Crude fat	4.18	2.18	2.38
Crude fiber	8.48	12.7	9.34
Ash	5.90	5.30	5.16
Calcium	0.79	0.76	0.81
Phosphorus	0.68	0.41	0.46
Sodium	0.25	0.15	0.13
Starch	37.4	34.6	37.0
Energy (MJ ME/kg DM)	13.31	11.54	13.15

The chemical composition of the two different feed ratios as well as of the restrictive dry feed is displayed in Table 2.

Sows in System II were identified by the transponder and automatically sorted by age and weight using an algorithm. With controlled access to low or high energy and nutrient feed, those sows were able to ingest feed ad libitum together at the cross trough at any time. If the target condition could be determined for a sow, it was assigned to the feeding area with lower-energy and -nutrient feed. The target condition or weight was set according to the BW of sows under restrictive feeding (Table 3).

Table 3. Sorting gate algorithm according to body weight (BW) and parity.

Parity	Target BW (kg)	Sorting to the High-Energy Ration BW (kg)	Sorting to the Low-Energy Ration BW (kg)
0	225	<215	>235
1 and 2	270	<260	>280
3 and 4	300	<290	>310
>5	325	<315	>335

2.4. Data Collection

2.4.1. Feed Intake

Daily feed consumption was calculated for both systems. For the sows in System II, the amount of daily feed allocation per valve was recorded by the computer of the liquid feeding system every day during the study. The number of sows using the feeding area was recorded by the sorting gate. From this, the average feed intake could be calculated. Feed losses at the trough could not be taken into account. The sows in System I received feed twice daily. The volumetric feeders installed in the compartment were gauged, and the feed quantities weighed. From this, the average feed intake per animal could be calculated.

2.4.2. Body Condition Development

The BW of the sows was determined on four control days always at the same time (entry into the waiting group, 70th gestation day, exit from the waiting group at 108th day of gestation, weaning) by means of mobile individual animal scales (T.E.L.L. Steuerungssysteme GmbH & Co.KG, Vreden, Germany; weighing range: 65–500 kg). To document the development of body condition, the body condition score (BCS) of all sows was assessed on the same four control days as the BW from the same single person, according to Kamphues et al. [24]. The BCS assessment was always done before determining the BW and back fat thickness (BFT). The BFT of the sows was measured using an ultrasonic device (Logiq V2, GE Healthcare, Inc., Wauwatosa, WI, USA) on the day of entry into the waiting group, at the exit from the waiting group, and at weaning. The measurement was done when the sows were standing with their backs straight on the individual animal scales using three measurement points on the right side of the body. The mean value of the three measurement points resulted in the BFT.

2.4.3. Health Status

In the peripartal period, manual or medical birth assistance, e.g., dystocia was recorded. In the puerperium, the body temperature of sows was measured directly after parturition and 12 h postpartum. After weaning, the weaning estrus interval was documented.

Vaginal injuries were documented in the waiting group. Clinical examination and evaluation of the external genitalia were performed on the 108th day of gestation, documenting any type of injury to the vulva at the end of gestation. Depending on the size of the injury, these were classified into three categories: (score 0) no injuries were found to the vulva; (score 1) fresh bloody injuries < 4 cm; (score 2) fresh bloody injuries > 4 cm.

2.4.4. Performance Data

At farrowing, the following reproductive parameters were recorded: litter number, number of piglets born alive, stillborn, mummified piglets, litter size of piglets born alive (within the first 12 h postpartum), number of piglets after litter balancing, as well as number of weaned piglets and litter size of weaned piglets. During the births, a total of 31 sows (16 in System I and 15 in System II) were filmed with camera systems. The exact length of birth (time interval from first to last born piglet), the birth interval between piglets (including mummified and stillborn piglets), and the vitality of the piglets could be documented for these sows. The following parameters were selected to describe piglet vitality immediately after birth according to a score developed by Muns et al. [25]: The piglets' ability to move was assessed based on body posture and position change (movement or rotation around body axis). Documented were head movements of piglets that corresponded to search behavior or actual mammary stimulation. The time interval from birth to first teat contact was recorded for each piglet.

2.5. Statistical Analyses

Statistical analysis was performed using the Statistical Analysis System for Windows, SAS®Enterprise Guide®, version 7.1 (SAS Institute Inc. Cary, NC, USA). Descriptive statistics were applied to report mean and standard deviation values. The Shapiro–Wilk test and Kolmogorov–Smirnov test were used to test the normal distribution of the data and verified by visual representation. If data could be considered normally distributed, a comparison was made using one-way ANOVA and the Ryan–Einot–Gabriel–Welsch test (probability of error α = 5%). In the case of not normally distributed data, the nonparametric Wilcoxon's two-sample test was applied. Correlation analyses were performed by determining Pearson's correlation coefficient. Indicated by SAS®, p-values with a significance level lower than α = 0.05 were considered statistically significant.

3. Results

3.1. Feed Intake

Daily feed intake of sows in the two feeding systems differed significantly ($p < 0.0001$). Both the diluted low-energy and concentrated high-energy rations in System II resulted in high feed intakes with very high variation in each case (low-energy ration 4.67 $\pm$ 2.14 kg dry matter (DM) per animal and day and high-energy ration 4.52 $\pm$ 2.03 kg DM per animal and day). Feed losses at the trough (during feed intake) were observed but could not be quantified. The sows in System I consumed 2.54 $\pm$ 0.23 kg DM per animal and day.

3.2. Body Condition Development

No significant differences in BW, BCS, and BFT were observed between the feeding systems during the gestation period. Interestingly, the mean BW, BCS, and BFT gain during gestation of the sows in System I were significantly higher (BW: 68.23 $\pm$ 24.92 vs. 57.03 $\pm$ 22.42, BCS: 0.54 $\pm$ 0.65 vs. 0.19 $\pm$ 0.48 and BFT: 4.30 $\pm$ 4.11 vs. 3.26 $\pm$ 6.49; $p < 0.05$, Table 4). However, at the beginning of the field study, sows in System I had a significantly lower BCS and BFT.

Table 4. Body condition development of sows during pregnancy.

Timepoint	BW (kg)		BCS		BFT (mm)	
	System I	System II	System I	System II	System I	System II
Entry into waiting group	225.43 $\pm$ 43.48 [a]	236.72 $\pm$ 59.84 [a]	3.25 $\pm$ 0.49 [a]	3.62 $\pm$ 0.60 [b]	11.49 $\pm$ 3.51 [a]	13.78 $\pm$ 4.56 [b]
Day 70 of gestation	268.86 $\pm$ 58.03 [a]	271.55 $\pm$ 57.49 [a]	3.63 $\pm$ 0.46 [a]	3.73 $\pm$ 0.50 [a]	-	-
Day 108 of gestation	295.34 $\pm$ 59.81 [a]	299.03 $\pm$ 61.31 [a]	3.79 $\pm$ 0.52 [a]	3.81 $\pm$ 0.55 [a]	15.84 $\pm$ 5.61 [a]	16.98 $\pm$ 5.66 [a]
Gain during gestation[3]	68.23 $\pm$ 24.92 [a]	57.03 $\pm$ 22.42 [b]	0.54 $\pm$ 0.65 [a]	0.19 $\pm$ 0.48 [b]	4.30 $\pm$ 4.11 [a]	3.26 $\pm$ 6.49 [b]

[a, b] different letters mark significant differences between means in a row for one parameter ($p < 0.05$).

3.3. Health Status of Sows

There were no significant differences between sows in the two feeding systems in the number of obstetric measures or in body temperature postpartum.

Significantly ($p < 0.05$) more vulva injuries at the end of gestation (day 108) were observed in System I (32.50% of sows (21.25% score 1 and 11.25% score 2) vs. 17.10% of sows (14.47% score 1 and 2.63% score 2) in System II).

3.4. Performance Data

There were no significant differences between the systems in terms of the overall performance of sows. In System I, significantly more piglets were left with a sow after litter balancing (Table 5). The crushing losses in System II were significantly reduced by one-third compared to System I. Recorded litter size of live-born piglets and calculated total litter size (with secundiae) were significantly different between feeding systems. Higher masses were recorded for sows in System II in each case. The calculated litter size of dead and live born piglets or the calculated average birth weight per piglet showed no significant differences. The individual BW of the piglets were determined at the time of weaning and were almost equal between the two systems.

Table 5. Birth and performance data of sows as well as litter sizes and weight of piglets at birth and weaning (kg).

Parameter	System I	System II
Piglets born alive per litter	16.71 [a] ± 5.32	18.49 [a] ± 3.52
Still born piglets per litter	1.80 [a] ± 1.97	1.47 [a] ± 1.83
Mummified born piglets per litter	1.01 [a] ± 2.06	0.79 [a] ± 1.42
Piglets after litter balancing	15.93 [a] ± 1.47	15.47 [b] ± 2.21
Crushed piglets per litter	1.20 [a] ± 0.91	0.80 [b] ± 0.80
Dead piglets (cause unknown) per litter	1.16 [a] ± 1.09	1.05 [a] ± 1.12
Weaned piglets per litter	12.75 [a] ± 1.58	12.75 [a] ± 1.40
Litter weight (dead and alive piglets)	23.84 [a] ± 6.11	25.55 [a] ± 4.18
Litter weight (piglets born alive) [1]	21.83 [a] ± 5.56	23.52 [b] ± 3.96
Average birth weight of piglets born alive [2]	1.24 [a] ± 0.22	1.31 [a] ± 0.23
Total litter size (with secundiae) [3]	28.16 [a] ± 8.02	30.61 [b] ± 5.01
Average weight of piglets at weaning	6.38 [a] ± 0.10	6.38 [a] ± 1.01

[a,b] Different letters indicate significant differences between the mean values in one row ($p < 0.05$). [1] Calculated assumption: same mean birth weight of piglets born alive and dead. [2] Calculated, from litter size and number of piglets born alive. [3] Calculated, as the sum of litter size and placental mass (placental mass = 19.8% of litter size [26]).

The sows (n =75) from System I could be inseminated after an average of 5.89 ± 5.35 days after weaning. Animals (n = 70) from System II had a comparable weaning estrus interval (5.87 ± 3.17 days).

3.5. Farrowing

For the duration of birth from first to last piglet, the individual time intervals between the birth of two piglets and the time until complete expulsion of the placenta after the last-born piglet, no significant differences were found between the sows of the two systems. Regarding the duration of birth and the mean birth interval, the sows in System II tended to take longer (06:49:45 ± 04:57:21 and 00:21:47 ± 00:47:22 in System I vs. 07:08:42 ± 04:35:33 and 00:24:24 ± 01:08:50 in System II), while the time interval between the birth of the last piglet until complete expulsion of placenta tended to be shorter (05:10:24 ± 06:02:55 in System I vs. 03:37:29 ± 03:08:30 in System II).

3.6. Piglet Vitality

The recorded video footage was examined for each piglet regarding the time interval until the first head and body movement and until the first contact with the teats of the respective piglet. The newborn piglets of sows from feeding System II took significantly less time to show head movement and to reach the teat (Table 6).

Table 6. Mean time interval until first body movement, first head movement, and first contact of piglets with teats [hh:mm:ss].

Time	System I	System II
Until body movement	00:01:16 [a] ± 00:01:06	00:01:13 [a] ± 00:01:08
Until head movement	00:00:31 [a] ± 00:00:47	00:00:27 [b] ± 00:00:36
Until contact with the teats	00:22:16 [a] ± 00:24:18	00:21:11 [b] ± 00:28:48

[a,b] Different letters mark significant differences between means in a row ($p < 0.05$).

Of the sows from System II, significantly ($p < 0.05$) more piglets (System I: 70.79%; System II: 81.15% of piglets) showed head movements analogous to mammary stimulation in the first 30 seconds after birth. In the first 10 minutes after birth, significantly more piglets from the sows in System II reached the teats in comparison to piglets from System I sows (43.35 vs. 32.79%).

3.7. Body Condition at Weaning

BW of the sows was recorded on the 108th day of gestation and on the day of weaning (day 28 postpartum). BW postpartum was calculated based on the determined litter size. The weight loss of sows from both systems during lactation and BW at weaning did not differ significantly from each other (Table 7). The display of the change in BW was only based on the animals for which the calculation of the weight postpartum was possible. At the end of gestation, the sows of the two feeding groups reached an almost identical mean BCS. The calculation of the mean BCS difference was based exclusively on the animals that had also been assessed on the 108th day of gestation. After lactation, there was no significant difference in the BCS values of the sow groups. During lactation, sows in feeding System II lost numerically more back fat; this difference was not significant.

Table 7. BW, BCS, and BFT change during lactation of the sows in both systems.

Timepoint	BW (kg)		BCS		BFT (mm)	
	System I	System II	System I	System II	System I	System II
Day 108 of gestation	295.34 [a] ± 59.81	299.03 [a] ± 61.31	3.79 [a] ± 0.52	3.81 [a] ± 0.55	15.84 [a] ± 5.61	16.98 [a] ± 5.66
Postpartum	265.67 [a] ± 56.44	267.49 [a] ± 61.64	-	-	-	-
Weaning	237.11 [a] ± 48.41	236.90 [a] ± 52.74	2.95 [a] ± 0.45	3.00 [a] ± 0.51	11.18 [a] ± 4.03	11.45 [a] ± 4.59
Difference	−28.20 [a] ± 19.17	−31.17 [a] ± 18.67	−0.83 [a] ± 0.38	−0.84 [a] ± 0.43	−4.38 [a] ± 3.29	−5.54 [a] ± 2.97

[a,b] Different letters mark significant differences between means in a row for one parameter ($p < 0.05$).

4. Discussion

The investigations in this field study within the "SWOF" project (sow welfare optimized feeding) aimed to compare two feeding systems for pregnant group-housed sows (restricted dry feeding, System I vs. ad libitum liquid feeding based on farm-grown forage, System II) in terms of body condition development during gestation, sow performance data and sow and piglet health in the peripartal period and lactation. System II was equipped with a sorting gate to give sows access to two different feeding areas with rations either high or low in energy and nutrients according to their body condition. In particular, this ad libitum system was expected to enable sows a natural feed intake rhythm with regard to their voluntary feed intake behavior as well as maintain the weight development, health status, and litter performance characteristics, and therefore improving overall animal welfare. In this study, data were collected on the farm from two systems diverging with regard to feeding access (restricted vs. ad libitum), feed composition ("conventional" vs. based on WPWS) and type (dry vs. liquid), group size (35–55 sows vs. 76–117 sows) and management (small stable vs. large dynamic group) with a total of 183 sows. Manifold factors have therefore to be considered when interpreting the results.

During pregnancy, the sows gained body mass regardless of the feeding system in the group. No significant differences could be detected at the three measurement time points during pregnancy. A comparable body condition development of sows during pregnancy between these two feeding concepts (restrictive vs. ad libitum) was not described by Hoy et al. [27], Steffens [28], or Ziron [29], who all found a higher body mass gain in the group of ad libitum fed sows. Feed consumption in System II was almost twice the amount of System I (low-energy ration 4.67 ± 2.14 kg DM per animal and day and high-energy ration 4.52 ± 2.03 kg DM per animal and day vs. 2.54 ± 0.23 kg DM per animal and day), but a major uncertainty in energy and nutrient supply that cannot be calculated remains the obvious feed losses in feeding areas of the ad libitum feeding system as the design of the troughs and the feeding technique was a prototype and has to be adjusted and improved regarding the use of fiber-rich liquid feed. The same extent of the standard deviation of BW over all measuring times is an indication that the sow groups did not grow apart depending on the ad libitum feeding system. A growing apart of the sow groups due to the different feed intake of the animals in an ad libitum feeding system was described by different authors [27,29,30], whereas Steffens [28] could not document any growing apart of the sows during pregnancy under ad libitum feeding when comparing the feeding systems. Peltoniemi et al. [31] found no significant difference in BW and BFT of sows at the beginning of their study, which investigated the effect of ad libitum feeding on body condition development and fertility in comparison with a conventionally fed group. After the third lactation, the ad libitum fed sows were significantly heavier and had a significantly higher mean BFT. Steffens [28] and Schade [30] also documented lower BFT at the beginning of gestation than after lactation in ad libitum fed sows. Consequently, the ad libitum fed sows could start the next pregnancy with a higher BFT. This finding could not be confirmed in the present study. Unfortunately, the sows could not be analyzed over several reproductive cycles, so that a statement about performance development is not possible. Several studies have shown that BW and BFT of sows at the end of gestation are directly related to ME and nutrient intake during gestation [32,33]. In the present study, the average energy consumption of sows in System I was higher than in System II (low and high-energy ration, Energy (MJ ME/kg DM); 13.31, 11.54, 13.15, respectively), which might be a possible reason for this result.

When looking at our results, considering both systems, a lower occurrence of vulva injuries was observed in System II at the end of gestation (day 108; 32.5% vs. 17.1% of sows in System I and II, respectively). A likely explanation, according to Rault et al. [34], is that pigs are a gregarious species, and when housed in groups, they establish dominance hierarchies. Consequently, group-housed sows are required to share resources and compete for feed [34]. Vulva biting is thereby a potent method of expelling sows without being in danger of being attacked in return [35]. Thus, the higher occurrence of vulva injuries found in sows in System I in our study might result from the feed restriction that could have been related to the competitive behavior in sows and which was the main cause of this injury [36]. According to Angermann et al. [36], the use of controlled entrance doors in the ad libitum liquid feeding system could explain the lower occurrence of vulva injuries as the automatic entrance door at the sorting gate was designed to protect the animals from being attacked by other sows.

In the present study, the reproductive performance was on a generally high level with litter sizes between 16.71 ± 5.32 and 18.49 ± 3.52 piglets born alive. Other recent studies reported lower litter sizes under different conditions: depending on different energy levels during late gestation, Rooney et al. [37] found litters with 13.5 to 15.3 piglets born alive, and Li et al. [38] fed different ratios of soluble to insoluble fiber and reported between 11.2 and 12.8 piglets born alive.

The reproductive performance characteristics, e.g., live-born, stillborn, mummified piglets as well as weaned piglets or the calculated average birth weight per piglet of sows, were not significantly influenced by the different feeding systems. Previous studies [39,40] as well reported no difference in the litter performance, the litter weight, or the number of born, born alive, stillborn, or mummified piglets between the restrictively vs. ad libitum fed group-housed pregnant sows. Van der Peet-Schwering et al. [41], for example, reported with 13.6 total born piglets in restricted feeding vs. 13.5 under ad libitum feeding conditions very similar results, and Martí et al. [42] reported 13.5 total born piglets with restricted feeding and 13.6 total born piglets with ad libitum feeding during late gestation. However, this is not in accordance with Cools et al. [43], who found a higher number of total born and born alive piglets to restrictively fed sows during the peripartal period compared to ad libitum fed sows (total born 11.5 vs. 10.5 piglets, live-born 11.1 vs. 10.2 piglets.

Compared to restricted dry feeding (System I), ad libitum liquid feeding (System II) led to a significantly higher litter size of live-born piglets and calculated total litter size (with secundiae) for sows in System II. This is in accordance with Li et al. [44], who reported an increased placental weight in sows fed a high-fiber diet. The crude fiber content in the dry compound diet fed to sows in System I was lower than in the feed in System II (low- and high-energy ration, crude fiber (% DM); 8.48, 12.7, and 9.34, respectively), which might be an explanation for our finding. Moreover, fiber-rich feeding of the sows during pregnancy has an effect on the ingesta passage (faster) and thus lowers the risk of prolonged farrowing due to constipation [19].

Interestingly, the farrowing duration and the birth interval were longer in sows from System II, while the time interval between the birth of the last piglet until complete expulsion of the placenta tended to be shorter (05:10:24 in System I vs. 03:37:29 in System II). However, in our study, due to technical reasons, the respective system could not be continued in the farrowing crates. Thus, at farrowing, the sows of both systems had already been fed the same feed for some days. A possible long-term effect of the respective feeding system can therefore be discussed but not be explained.

It has been observed that a longer farrowing duration is associated with an increased piglet stillborn rate [45,46]. However, there was no evidence for a difference in the stillborn rate in this study, regardless of the feeding system. In contrast, Feyera et al. [47], who observed an influence of the timing of the last meal prior to parturition on the farrowing duration and stillborn rate, concluded that when sows had been offered a meal less than 6 h before parturition, sows had a shorter farrowing duration and a reduced stillborn rate.

With increasing litter size and performance selection, the piglet vitality or the viability of the piglets is also a relevant criterion for piglet production [48]. Birth weight is often a decisive factor for assessing viability and selection with regard to litter compensation [49,50]. In the study by Muns et al. [25], a relation was found between the ability to move and reach the teats of sows to assess piglet vitality and its influence on piglet survival and growth during lactation. In the present study, the newborn piglets from feeding System II showed smaller time intervals until first head movements and reaching the teats of the mother sows. The head movements analogous to mammary stimulation were a reliable indicator of the piglets' vitality and were correlated with their BW [25]. After evaluating the recorded video material, more piglets from the sows from System II showed head movements analogous to mammary stimulation in the first 30 seconds after birth (70.8% vs. 81.2%), and in the first ten minutes after birth, significantly more piglets from System II sows reached the teats in comparison to piglets from System I sows (43.4 vs. 32.8%). According to the evaluations by Muns et al. [25], mammary stimulation is a reliable indicator of piglet vitality. Nonetheless, ad libitum liquid feeding did not affect lactating and weaning sow performance. In accordance with van der Peet-Schwering et al. [41], it is possible to feed group-housed gestating sows a liquid diet ad libitum without negative effects on reproductive performance.

Unfortunately, due to technical reasons, the respective feeding system could not be continued during lactation, which could also have had effects on performance and health [43]. When using an ad libitum feeding system based on whole plant silages, higher feces masses are produced in combination with lower digestibility. These should be taken into account when designing stables and optimizing manure management [51]. Further studies are, therefore, ultimately necessary and useful to examine the effects of forage-based ad libitum feeding in an optimized housing system, since in terms of comprehensive animal welfare, the target is full feeding of sows.

5. Conclusions

In summary, it can be said that the tested group-adapted ad libitum feeding system in combination with a sorting gate had no negative impact on the body condition development and sow's performance. If anything, the piglet vitality was positively affected. The feeding concept evaluated in this study offers opportunities for more animal welfare, and due to the use of farm-grown forage, it also enhances sustainability and farm self-sufficiency. Nevertheless, this will require further research in the future regarding the repertoire of possibly applied farm-grown forage, the effects of the concept in the transit phase of sows, and the optimization of floor design and manure management.

Author Contributions: Conceptualization, C.V., F.v.u.z.M. and J.K.; methodology, C.V., J.K. and S.F.; software/hardware, R.H; validation, B.C., C.B.H., C.V. and S.F.; formal analysis, S.F. and C.V.; investigation, C.B.H., E.A., P.E. and S.F.; resources, C.V., R.H. and S.F.; data curation, B.C., C.B.H., C.V. and S.F.; writing—original draft preparation, C.B.H.; writing—review and editing, B.C., C.B.H., C.V., E.A., F.v.u.z.M., J.K., P.E., R.H. and S.F.; visualization, B.C. and C.B.H.; supervision, C.V.; project administration, C.V.; funding acquisition, C.V. and J.K. All authors have read and agreed to the published version of the manuscript.

Funding: The project was supported by funds of the Federal Ministry of Food and Agriculture (BMEL) based on a decision of the Parliament of the Federal Republic of Germany via the Federal Office for Agriculture and Food (BLE) under the innovation support program (Funding code 28RZ-3-72.063/745 625). This publication was supported by Deutsche Forschungsgemeinschaft and University of Veterinary Medicine Hannover Foundation within the funding program Open Access Publishing.

Institutional Review Board Statement: The study design was approved by the Animal Welfare Officer of the University of Veterinary Medicine Hannover, Foundation, Hannover, reference: TVO-2017-V-49.

Informed Consent Statement: Not applicable.

Data Availability Statement: The original contributions generated for the study are included in the article; further inquiries can be directed to the corresponding author.

Conflicts of Interest: The authors declare no conflict of interest. The funders had no role in the design of the study; in the collection, analyses, or interpretation of data; in the writing of the manuscript, or in the decision to publish the results.

References

1. Fraser, D. Understanding animal welfare. *Acta Vet. Scand.* **2008**, *50*, S1. [CrossRef] [PubMed]
2. Sandøe, P.; Hansen, H.O.; Rhode, H.L.H.; Houe, H.; Palmer, C.; Forkman, B.; Christensen, T. Benchmarking farm animal welfare—A novel tool for cross-country comparison applied to pig production and pork consumption. *Animals* **2020**, *10*, 955. [CrossRef] [PubMed]
3. Kemper, N. Update on postpartum dysgalactia syndrome in sows. *J. Anim. Sci.* **2020**, *98*, S117–S125. [CrossRef] [PubMed]
4. Oliviero, C.; Heinonen, M.; Valros, A.; Peltoniemi, O. Environmental and sow-related factors affecting the duration of farrowing. *Anim. Reprod. Sci.* **2010**, *119*, 85–91. [CrossRef]
5. Hughes, P.; Smits, R.; Xie, Y.; Kirkwood, R. Relationships among gilt and sow live weight, P2 backfat depth, and culling rates. *J. Swine Health Prod.* **2010**, *18*, 301–305.
6. Schenkel, A.; Bernardi, M.; Bortolozzo, F.; Wentz, I. Body reserve mobilization during lactation in first parity sows and its effect on second litter size. *Livest. Sci.* **2010**, *132*, 165–172. [CrossRef]
7. Bench, C.; Rioja-Lang, F.; Hayne, S.; Gonyou, H. Group gestation sow housing with individual feeding—II: How space allowance, group size and composition, and flooring affect sow welfare. *Livest. Sci.* **2013**, *152*, 218–227. [CrossRef]
8. Woods, A. Rethinking the history of modern agriculture: British pig production, c. 1910–65. *Twent. Century Br. Hist.* **2012**, *23*, 165–191. [CrossRef]
9. Stein, H.; Lagos, L.; Casas, G. Nutritional value of feed ingredients of plant origin fed to pigs. *Anim. Feed Sci. Technol.* **2016**, *218*, 33–69. [CrossRef]
10. Werner, C.; Sundrum, A. Zum Einsatz von Raufutter bei Mastschweinen. *Landbauforsch. Völkenrode Sonderh.* **2008**, *320*, 61–68.

11. Bergeron, R.; Bolduc, J.; Ramonet, Y.; Meunier-Salaün, M.; Robert, S. Feeding motivation and stereotypies in pregnant sows fed increasing levels of fibre and/or food. *Appl. Anim. Behav. Sci.* **2000**, *70*, 27–40. [CrossRef]
12. Kallabis, K.E.; Kaufmann, O. Effect of a high-fibre diet on the feeding behaviour of fattening pigs. *Arch. Anim. Breed.* **2012**, *55*, 272–284. [CrossRef]
13. Read, E.; Baxter, E.; Farish, M.; D'Eath, R. Trough half empty–pregnant sows are fed under half of their ad libitum intake. *Anim. Welfare* **2020**, *29*, 151–162. [CrossRef]
14. D'Eath, R.B.; Jarvis, S.; Baxter, E.M.; Houdijk, J. Mitigating hunger in pregnant sows. In *Advances in Pig Welfare*; Špinka, M., Ed.; Woodhead Publishing: Cambridge, UK, 2018; pp. 199–234.
15. Bernardino, T.; Tatemoto, P.; de Moraes, J.E.; Morrone, B.; Zanella, A.J. High fiber diet reduces stereotypic behavior of gilts but does not affect offspring performance. *Appl. Anim. Behav. Sci.* **2021**, *243*, 105433. [CrossRef]
16. Priester, M.; Visscher, C.; Fels, M.; Rohn, K.; Dusel, G. Fibre supply for breeding sows and its effects on social behaviour in group-housed sows and performance during lactation. *Porcine Health Manag.* **2020**, *6*, 15. [CrossRef] [PubMed]
17. Mou, D.; Li, S.; Yan, C.; Zhang, Q.; Li, J.; Wu, Q.; Qiu, P.; He, Y.; Li, Y.; Liu, H. Dietary fiber sources for gestation sows: Evaluations based on combined in vitro and in vivo methodology. *Anim. Feed Sci. Technol.* **2020**, *269*, 114636. [CrossRef]
18. Presto, M.; Rundgren, M.; Wallenbeck, A. Inclusion of grass/clover silage in the diet of growing/finishing pigs–Influence on pig time budgets and social behaviour. *Acta Agric. Scand. A Anim. Sci.* **2013**, *63*, 84–92. [CrossRef]
19. Kleine, S.U. Untersuchungen zum Einfluss von Maissilage als Rohfaserquelle in der Flüssigfütterung Tragender Sauen auf Gesundheit und Leistung in der Folgenden Laktation. Ph.D. Thesis, University of Veterinary Medicine Hannover, Hanover, Germany, 2012.
20. De Leeuw, J.; Bolhuis, J.; Bosch, G.; Gerrits, W. Effects of dietary fibre on behaviour and satiety in pigs: Symposium on 'Behavioural nutrition and energy balance in the young'. *Proc. Nutr. Soc.* **2008**, *67*, 334–342. [CrossRef]
21. Aubé, L.; Guay, F.; Bergeron, R.; Bélanger, G.; Tremblay, G.; Devillers, N. Sows' preferences for different forage mixtures offered as fresh or dry forage in relation to botanical and chemical composition. *Animal* **2019**, *13*, 2885–2895. [CrossRef]
22. Havet, A.; Coquil, X.; Fiorelli, J.-L.; Gibon, A.; Martel, G.; Roche, B.; Ryschawy, J.; Schaller, N.; Dedieu, B. Review of livestock farmer adaptations to increase forages in crop rotations in western France. *Agric. Ecosyst. Environ.* **2014**, *190*, 120–127. [CrossRef]
23. Kambashi, B.; Boudry, C.; Picron, P.; Bindelle, J. Forage plants as an alternative feed resource for sustainable pig production in the tropics: A review. *Animal* **2014**, *8*, 1298–1311. [CrossRef] [PubMed]
24. Kamphues, J.; Wolf, P.; Coenen, M.; Eder, K.; Iben, C.; Kienzle, E.; Liesegang, A.; Männer, K.; Zebeli, Q.; Zentek, J. *Supplemente zur Tierernährung für Studium und Praxis*; Schlütersche: Hanover, Germany, 2014.
25. Muns, R.; Manzanilla, E.G.; Sol, C.; Manteca, X.; Gasa, J. Piglet behavior as a measure of vitality and its influence on piglet survival and growth during lactation. *J. Anim. Sci.* **2013**, *91*, 1838–1843. [CrossRef]
26. Antonczyk, C. *Untersuchungen zu Energie-und Nährstoffgehalten in der Frucht-und Plazentamasse zum Zeitpunkt der Geburt bei Sauen mit Hoher Reproduktionsleistung*; DVG Service GmbH: Giessen, Germany, 2017.
27. Hoy, S.; Ziron, M.; Leonhard, P.; Oppong Sefa, K. Untersuchungen zum Futteraufnahmeverhalten ad libitum gefütterter tragender Sauen in Gruppenhaltung an Rohrautomaten. *Arch. Anim. Breed.* **2001**, *44*, 629–638. [CrossRef]
28. Steffens, R. Gesundheit und Leistung von Sauen nach Unterschiedlicher Fütterung in der Trächtigkeit (Übliches Alleinfutter Restriktiv im Vergleich zu Trockenschnitzelreichem Mischfutter ad Libitum). Ph.D. Thesis, University of Veterinary Medicine Hannover, Hanover, Germany, 2005.
29. Ziron, M. *Einfluss der ad Libitum bzw. Rationierten Fütterung von Sauen über Mehrere Trächtigkeiten Hinweg auf Unterschiedliche Verhaltens-und Leistungsparameter*; Universität Gießen: Gießen, Germany, 2005.
30. Schade, C. Feldstudie zu Auswirkungen Einer ad-Libitum-Fütterung in der Gravidität auf die Entwicklung von Futteraufnahme, Körpermasse, Ernährungszustand und Rückenspeckdicke Sowie auf die Reproduktionsleistung und Gesundheit von Sauen. Ph.D. Thesis, University of Veterinary Medicine Hannover, Hanover, Germany, 2000.
31. Peltoniemi, O.; Tast, A.; Heinonen, M.; Oravainen, J.; Munsterhjelm, C.; Hälli, O.; Oliviero, C.; Hämeenoja, P.; Virolainen, J.V. Fertility of sows fed ad libitum with a high fibre diet during pregnancy. *Reprod. Domes. Anim.* **2010**, *45*, 1008–1014. [CrossRef] [PubMed]
32. Dourmad, J.-Y. Effect of feeding level in the gilt during pregnancy on voluntary feed intake during lactation and changes in body composition during gestation and lactation. *Livest. Prod. Sci.* **1991**, *27*, 309–319. [CrossRef]
33. Long, H.; Ju, W.; Piao, L.; Kim, Y. Effect of dietary energy levels of gestating sows on physiological parameters and reproductive performance. *Asian Australas. J. Anim. Sci.* **2010**, *23*, 1080–1088. [CrossRef]
34. Rault, J.; Ho, H.; Verdon, M.; Hemsworth, P. Feeding behaviour, aggression and dominance in group-housed sows. *Anim. Prod. Sci.* **2015**, *55*, 1495. [CrossRef]
35. Van Putten, G.; Van de Burgwal, J. Vulva biting in group-housed sows: Preliminary report. *Appl. Anim. Behav. Sci.* **1990**, *26*, 181–186. [CrossRef]
36. Angermann, E.; Raoult, C.; Wensch-Dorendorf, M.; Frenking, S.; Kemper, N.; von Borell, E. Development of a Group-Adapted Housing System for Pregnant Sows: A Field Study on Performance and Welfare Aspects. *Agriculture* **2021**, *11*, 28. [CrossRef]
37. Rooney, H.B.; O'driscoll, K.; O'doherty, J.V.; Lawlor, P.G. Effect of increasing dietary energy density during late gestation and lactation on sow performance, piglet vitality, and lifetime growth of offspring. *J. Anim. Sci.* **2020**, *98*, skz379. [CrossRef]

38. Li, Y.; Zhang, L.; Liu, H.; Yang, Y.; He, J.; Cao, M.; Yang, M.; Zhong, W.; Lin, Y.; Zhuo, Y. Effects of the ratio of insoluble fiber to soluble fiber in gestation diets on sow performance and offspring intestinal development. *Animals* **2019**, *9*, 422. [CrossRef]
39. Petherick, J.C.; Blackshaw, J.K. A note on the effect of feeding regime on the performance of sows housed in a novel group-housing system. *Anim. Sci.* **1989**, *49*, 523–526. [CrossRef]
40. Whittaker, X.; Edwards, S.; Spoolder, H.; Corning, S.; Lawrence, A. The performance of group-housed sows offered a high fibre diet ad libitum. *Anim. Sci.* **2000**, *70*, 85–93. [CrossRef]
41. Van der Peet-Schwering, C.; Kemp, B.; Plagge, J.; Vereijken, P.; Den Hartog, L.; Spoolder, H.; Verstegen, M. Performance and individual feed intake characteristics of group-housed sows fed a nonstarch polysaccharides diet ad libitum during gestation over three parities. *J. Anim. Sci.* **2004**, *82*, 1246–1257. [CrossRef]
42. Martí, L.; Latorre, M.Á.; Álvarez-Rodríguez, J. Does ad libitum feeding during the peri-partum improve the sow feed intake and performances? *Animals* **2019**, *9*, 1078. [CrossRef] [PubMed]
43. Cools, A.; Maes, D.; Decaluwé, R.; Buyse, J.; van Kempen, T.A.; Liesegang, A.; Janssens, G. Ad libitum feeding during the peripartal period affects body condition, reproduction results and metabolism of sows. *Anim. Reprod. Sci.* **2014**, *145*, 130–140. [CrossRef] [PubMed]
44. Li, Y.; He, J.; Zhang, L.; Liu, H.; Cao, M.; Lin, Y.; Xu, S.; Fang, Z.; Che, L.; Feng, B.; et al. Effects of dietary fiber supplementation in gestation diets on sow performance, physiology and milk composition for successive three parities. *Anim. Feed Sci. Technol.* **2021**, *276*, 114945. [CrossRef]
45. Gourley, K.M.; Swanson, A.J.; Royall, R.Q.; DeRouchey, J.M.; Tokach, M.D.; Dritz, S.S.; Goodband, R.D.; Hastad, C.W.; Woodworth, J.C. Effects of timing and size of meals prior to farrowing on sow and litter performance. *Transl. Anim. Sci.* **2020**, *4*, 724–736. [CrossRef]
46. Langendijk, P.; Plush, K. Parturition and its relationship with stillbirths and asphyxiated piglets. *Animals* **2019**, *9*, 885. [CrossRef]
47. Feyera, T.; Pedersen, T.F.; Krogh, U.; Foldager, L.; Theil, P.K. Impact of sow energy status during farrowing on farrowing kinetics, frequency of stillborn piglets, and farrowing assistance. *J. Anim. Sci.* **2018**, *96*, 2320–2331. [CrossRef]
48. Kemper, N. Was tun mit großen Würfen–sind wir machtlos? In Proceedings of the Leipz. Blaue Hefte, 8. Leipziger Tierärztekongress, Conference Transcript, Leipzig, Germany, November 2015; Volume 3, p. 157.
49. Milligan, B.N.; Fraser, D.; Kramer, D.L. Within-litter birth weight variation in the domestic pig and its relation to pre-weaning survival, weight gain, and variation in weaning weights. *Livest. Prod. Sci.* **2002**, *76*, 181–191. [CrossRef]
50. Deen, M.; Bilkei, G. Cross fostering of low-birthweight piglets. *Livest. Prod. Sci.* **2004**, *90*, 279–284. [CrossRef]
51. Ebertz, P.; Schmithausen, A.J.; Büscher, W. Ad libitum feeding of sows with whole crop maize silage—Effects on slurry parameters, technology and floor pollution. *Anim. Feed Sci. Technol.* **2020**, *262*, 114368. [CrossRef]

 *sustainability*

MDPI

Article

Commercial Corn Hybrids as a Single Source of Dietary Carotenoids: Effect on Egg Yolk Carotenoid Profile and Pigmentation

Kristina Kljak, Marija Duvnjak, Dalibor Bedeković, Goran Kiš *, Zlatko Janječić and Darko Grbeša

Faculty of Agriculture, University of Zagreb, Svetošimunska Cesta 25, 10000 Zagreb, Croatia; kkljak@agr.hr (K.K.); mduvnjak@agr.hr (M.D.); dbedekovic@agr.hr (D.B.); zjanjecic@agr.hr (Z.J.); dgrbesa@agr.hr (D.G.)
* Correspondence: kis@agr.hr; Tel.: +385-1-239-3933

Abstract: Commercial high-yielding corn hybrids have not been evaluated for their ability to pigment egg yolk. Therefore, the objective of this research was to investigate the effects of commercial hybrids with different carotenoid profiles as the only source of pigments in the diets of hens on yolk color and carotenoid content, as well as the carotenoid deposition efficiency into the yolk. Treatment diets, differing only in one of five corn hybrids, were offered in a completely randomized design in six cages per treatment, each with three hens. Treatment diets and yolks differed in carotenoid profile (contents of lutein, zeaxanthin, β-cryptoxanthin and β-carotene, $p < 0.001$), with total carotenoid contents ranging from 17.13–13.45 µg/g in diet and 25.99–21.97 µg/g in yolk. The treatments differed ($p < 0.001$) in yolk color, which was determined by yolk color fan (10.8–9.83) and CIE Lab (redness; range 12.47–10.05). The highest yolk color intensity was achieved by a diet with the highest content of zeaxanthin, β-cryptoxanthin and β-carotene. The deposition efficiency of lutein and zeaxanthin (25.52 and 26.05%, respectively) was higher than that of β-cryptoxanthin and β-carotene (8.30 and 5.65%, respectively), and the deposition efficiency of all carotenoids decreased with increasing dietary content. Commercial corn hybrids provided adequate yolk color and could be the only source of carotenoids in the diets of hens, which could reduce the cost of egg production and increase farmers' income.

Keywords: corn hybrid; egg yolk; lutein; zeaxanthin; β-cryptoxanthin; β-carotene; yolk color; carotenoid deposition efficiency

Citation: Kljak, K.; Duvnjak, M.; Bedeković, D.; Kiš, G.; Janječić, Z.; Grbeša, D. Commercial Corn Hybrids as a Single Source of Dietary Carotenoids: Effect on Egg Yolk Carotenoid Profile and Pigmentation. *Sustainability* **2021**, *13*, 12287. https://doi.org/10.3390/su132112287

Academic Editors: Nikola Puvača, Vincenzo Tufarelli and Eva Voslarova

Received: 29 September 2021
Accepted: 2 November 2021
Published: 7 November 2021

Publisher's Note: MDPI stays neutral with regard to jurisdictional claims in published maps and institutional affiliations.

1. Introduction

Producing and providing an adequate amount of high-quality food for the rapidly growing world population requires enormous resources that negatively impact the environment [1]. Among other environmental protection methods, the use of feeds that meet more than one animal requirement shows promising potential by reducing the impact of the production of additional ingredients (such as vitamins and pigments) and contributing to the reduction in the environmental footprint of animal feed production. In this regard, corn grain has the potential to meet poultry energy and resistant starch requirements, as well as contribute to the antioxidant capacity and pigmentation of egg yolk.

The natural color of the yolk is a result of accumulated carotenoids due to hens' ability to deposit carotenoids in the yolk. Carotenoids are compounds with pigmenting but also antioxidant and provitamin A activity, which is why higher yolk color intensity is associated with better laying hen health and better egg quality and flavor [2]. Eggs, in turn, are important natural sources of carotenoids in the human diet, with the xanthophylls lutein and zeaxanthin being prominent due to the prevention and reduction of cataracts and age-related macular degeneration [3,4]. Similar to other animals, hens cannot synthesize carotenoids and must obtain them through diet.

Yolk color is one of the most important attributes for consumer preference in eggs, and one of the most commonly used methods for determining yolk color intensity is the DSM Yolk Color Fan scale (YCF). Yolk color preference varies from country to country, and consumers in Southern Europe prefer a more intense yolk color (YCF > 11), while the majority of consumers worldwide prefer paler eggs [2]. To achieve the desired yolk color, carotenoids are supplied in hen diets as pigment additives, which is an important contribution to the overall cost of egg production. The global carotenoids market reached USD 1.5 billion in 2017 and is expected to reach USD 2.0 billion by 2022, growing at a compound annual growth rate of 5.7% from 2017 to 2022 [5].

A number of factors can influence yolk color, but when considering carotenoids, their content, type and ratio in the diet presents the basis for the possible color [6,7]. In general, the carotenoid profile of the yolk reflects the carotenoid profile of the diet [8], and consequently, the yolk color can be influenced by the addition of different carotenoid sources in the hen diet. It is considered that the standard hen diet, in which the main part is one or more cereals, is insufficient to achieve the desired yolk color without pigment supplementation. These pigments are usually of synthetic origin as they deposit faster and have better efficiency than natural pigments [2,9]. However, when such pigments are added to hen diets, the higher yolk color intensity is not associated with the higher yolk carotenoid content [10]. Furthermore, with the increasing concern regarding healthy nutrition, the synthetic supplementation of food and feed has become an issue. Although numerous natural sources have been investigated over the last three decades, only marigold is commercially available [11].

Corn, which is often the main ingredient of hen diets, is the only cereal with appreciable carotenoid content and an adequate carotenoid profile for egg yolk pigmentation [12]. However, the contribution of corn carotenoids to yolk pigmentation is neglected even in countries with high corn content in hen diets, and pigment supplementation is common in corn-soy-based diets, despite increase of costs and decrease of carotenoid deposition efficiency [6]. In the last decade, biofortification has been used to create corn varieties with enhanced carotenoid content, mainly for human nutrition [13]. When biofortified corn is the only source of carotenoids in hen diets, eggs have higher carotenoid content and more intensive color of yolks compared to non-biofortified corn [14,15]. However, corn varieties biofortified with carotenoids have a much lower yield and a higher price compared to commercial corn hybrids [16], which could limit their use in intensive crop and poultry production.

The most important reason for underestimating the contribution of corn carotenoids to yolk color is likely due to the high genetic variability of carotenoid content in high-yielding hybrids used in commercial production. This genetic variability of corn hybrids suggests that hybrids with high total carotenoid content and adequate profile for yolk pigmentation can be found in the market. At the same time, the use of such hybrids in hen diets does not require pigment addition, which simplifies diet preparation, reduces egg production costs and increases farmers' income. However, commercially available high-yielding corn hybrids have not been evaluated for their ability to pigment the yolk and the efficiency of carotenoid deposition. Therefore, the objective of this research was to investigate the effects of commercial high-yielding hybrids with different carotenoid profiles as the only source of hen diet pigment on yolk color and carotenoid content, and carotenoid deposition efficiency into yolk.

2. Materials and Methods

The animal experiment was conducted in accordance with the Croatian directives (Animal Protection Act, OG 102/17, and Regulation on the Protection of Animals Used for Scientific Purposes, OG 55/13), which correspond to the European guidelines for the care and use of animals used for scientific purposes. The animal procedures used in this study were approved by the Bioethics Committee for the protection and welfare of animals at the University of Zagreb Faculty of Agriculture (KLASA 114-04/17-03/02, URBROJ 251-71-01-17-1).

2.1. Corn Hybrids and Treatment Diets

Five commercial and high-yielding yellow corn hybrids (Zea Mays L.) with different carotenoid content (Bc 572, Kekec, Mejaš, Riđan and Pajdaš) were provided by Bc Institute (Zagreb, Croatia). The corn hybrids were grown under the same agro-climate and production conditions. Each hybrid was planted on a 560 m^2 test plot located in central Croatia, near Zagreb. At the time of physiological maturity of the grain, each corn hybrid was harvested from the central part of the plot. The harvested grains were dried at 60 °C to 120 g/kg moisture and stored in corn storage bags until the preparation of the experimental diets.

The five experimental diets were formulated to have the same nutrient content and differed only in the corn hybrid. All diets were formulated according to the NRC [17] and updated to be suitable for TETRA SL commercial layer hen hybrid at initial phase of egg production (20–45 weeks of age) [18]; the diet composition and calculated nutrient contents are shown in Table 1. No pigment source other than corn was added. Immediately prior to preparation of the diets, corn grain was ground to pass through a 4-mm screen. Since the objective of the study was to compare commercial high-yielding corn hybrids, the control diet was not included in the experiment.

Table 1. Diet composition and calculated nutrient content.

Ingredient	Content (g/kg)
Corn	600
Soybean meal	262
Sunflower oil	30
Calcium carbonate	88
Monocalcium phosphate	12
Sodium chloride	4
DL methionine	1.5
Vitamin premix [1]	1.2
TRT Poultry Pack [2]	1.3
Calculated nutrient composition	
Crude protein	170
Crude fat	55
Crude fibre	28
Crude ash	12.7
Calcium	38
Phosphorus, available	4.3
Lysine	8.8
Methionine	4.2
Metabolic energy (MJ kg^{-1})	11.6

[1] The vitamin premix provided per kg of diet: Vitamin A 10,000 IU, Vitamin D3 2500 IU, Vitamin E 200 mg, Vitamin K3 3 mg, Vitamin B1 1 mg, Vitamin B2 45 mg, Vitamin B3 30 mg, Vitamin B5 10 mg, Vitamin B6 3 mg, Vitamin B7 50 mg, Vitamin B9 0.5 mg, Vitamin B12 25 mg, Choline 400 mg, antioxidant (BHA, EQ) 50 mg. [2] TRT Poultry Pack (Alltech Ireland Ltd., Dunboyne, Ireland) provided per kg of diet: I 1 mg, Fe 5 mg, Cu 5 mg, Mn 30 mg, Zn 30 mg, Se 0.2 mg.

All diets were mixed immediately before the start of the dietary experiment and divided into 5 paper bags. For further analysis, a sample was taken from each bag of the same treatment diet (total of 5 replicates per treatment diet), and samples were stored at −20 °C until carotenoid analysis. Before analysis, corn samples were ground in a laboratory mill (Cyclotec 1093, Foss Tocator, Hoganas, Sweden) equipped with a 0.3-mm screen. All samples were analyzed for dry matter content (DM) determined by drying 3 g of each sample 4 h at 103 °C according to the method ISO 6496:1999 [19].

2.2. Hens, Housing and Experimental Design

A total of 90 TETRA-SL 18-week-old laying hens were randomly allotted in groups of 3 to 1 of 30 metal battery cages with 750 cm^2 per hen. Diets and water were provided ad libitum to hens. Room temperature was 20 ± 3 °C, and the light period consisted of 16 h light per day throughout the experimental period.

After allocating hens to the cages, a 4-week depletion period began. All hens were fed a diet without added pigments and based on barley instead of corn grain and with the same calculated nutrient composition as the experimental diets (Table 1). After depletion, the cages were randomly assigned to one of five dietary treatments (six replicates per dietary treatment). The experimental period lasted 10 weeks. Throughout the experimental period, the number of eggs laid was recorded daily, and diet intake was recorded weekly.

During the experimental period, eggs were collected every three days until the third week to determine stabilization in carotenoid content (i.e., on day 1, 4, 7, 10, 13, 16, 19 and 22 after the beginning of experimental period) and then weekly for color and carotenoid analysis (i.e., on day 28, 35, 42, 49, 56, 63 and 70 after the beginning of experimental period). All eggs were analyzed in the shortest possible time and stored at 4 °C if necessary. Collected eggs were broken immediately before analysis; the yolks were separated from albumen and dried on a paper napkin. After the yolks were separated, their weight was recorded.

2.3. Yolk Color Determination

Yolk color was analyzed with both YCF (Egg Multi Tester EMT-5200, Robotmation Co. Ltd., Japan) and Minolta Chroma Meter CR-410 (Minolta Co. Ltd., Osaka, Japan) using the CIE (Commission Internationale d'Eclairage) Lab scale. Color was first determined using the YCF scale; each yolk from eggs collected in the same cage on the same day was analyzed separately, and the average value was taken. Then, the yolks from each cage were combined and mixed carefully to avoid air bubbles, and the color was determined using the CIE Lab scale. The L*, a* and b* values reflect brightness (0 = black, 100 = white), redness (−a = green, a = red) and yellowness (−b = blue, b = yellow), respectively.

2.4. Carotenoid Analysis

2.4.1. Experimental Diets

Carotenoids from experimental diets were extracted and quantified according to the procedure described by Kurilich and Juvik [20], using β-apo-carotenal as an internal standard. Each sample was analyzed in triplicate, and the average value was taken as the result. For the extraction, samples were homogenized with ethanol, saponified with 80% KOH and incubated for 10 min at 85 °C in a water bath. The test tubes were then cooled in an ice bath with the addition of deionized water. The carotenoids were extracted with hexane, which was pipetted into a separate tube after centrifugation at 2200× *g* for 10 min (Centric 322A, Tehtnica, Železniki, Slovenia). The extraction procedure was repeated until the colorless upper hexane layer. The collected supernatants were evaporated using vacuum evaporator (Laborata 400 efficient, Heidolph, Schwabach, Germany) and reconstituted in 200 µL acetonitrile:dichloromethane:methanol (45:20:35, *v/v/v*) containing 0.1% BHT.

Carotenoids were separated and quantified using a SpectraSystem HPLC instrument (Thermo Separation Products, Inc., Waltham, MA, USA) equipped with a quaternary gradient pump, an autosampler and a UV/Vis detector. Two sequentially connected C18 reversed-phase columns, Vydac 201TP54 column (5 µm, 4.6 × 150 mm; Hichrom, Reading, UK) and Zorbax RX-C18 column (5 µm, 4.6 × 150 mm; Agilent Technologies, Santa Clara, CA, USA), were used for carotenoid separation. The separation columns were protected by a Supelguard Discovery C18 guard column (5 µm, 4 × 20 mm; Supelco, Bellefonte, PA, USA). The mobile phase consisted of acetonitrile:methanol:dichloromethane (75:25:5, *v/v/v*) containing 0.1% BHT and 0.05% triethylamine. An aliquot of 30 µL was injected, and the flow rate was 1.8 mL/min. The separations were performed at room temperature, and carotenoids were monitored at 450 nm.

Carotenoids (lutein (purity 99%), zeaxanthin (purity 99%), β-cryptoxanthin (purity 99%) and β-carotene (purity 98%)) were identified by comparing their retention times and quantified by external standardization with calibration curves using commercially available standards (Sigma-Aldrich, Steinheim, Germany; $r^2 \geq 0.99$ for all carotenoids). The total carotenoid content was calculated by summing the contents of the individual carotenoids.

2.4.2. Yolks

The changes in carotenoid content in yolks during the experimental period were determined using the reversed-phase HPLC method described previously (Section 2.4.1). Yolks mixed after YCF color analysis were weighed for the carotenoid extraction procedure described by Surai et al. [21] before the determination of color by CIE Lab. Carotenoids were extracted with hexane after yolk homogenization with 2 mL of 5% NaCl aq. solution:ethanol (1:1, *v/v*). Combined extracts were evaporated using a vacuum evaporator and reconstituted in 300 μL acetonitrile:dichloromethane:methanol (45:20:35, *v/v/v*) containing 0.1% BHT.

2.5. Carotenoid Deposition Efficiency

The carotenoid deposition efficiency for each cage within dietary treatment was calculated using the following equation [8]:

$$\text{Carotenoid deposition efficiency (\%) = Carotenoid production by egg / Carotenoid consumption by diet} \times 100$$

where carotenoid production by eggs and consumption by diet were calculated using the following equations:

$$\text{Carotenoid production by egg = yolk weight (g)} \times \text{yolk carotenoid content (μg/g)} \times \text{egg production (\%)}$$

$$\text{Carotenoid consumption by diet = diet intake (g/d/hen)} \times \text{diet carotenoid content (μg/g)}$$

based on the data obtained in the hen trial and after sample analysis.

2.6. Statistical Analysis

Statistical analyses of the obtained results were performed using SAS statistical software (version 9.4; SAS Institute Inc., Cary, NC, USA). The dietary experiment was conducted as a randomized block design with five dietary treatments, defining a cage with three hens as the experimental unit. Differences between the treatment diets were subjected to an analysis of variance using the MIXED procedure with treatment as the fixed effect. The same procedure was used to analyze differences between treatments in yolk color, carotenoid content and carotenoid deposition efficiency using repeated measurements ANOVA, with results obtained from third week and until end of the dietary experiment. Mean values were defined by the least squares means statement and compared using the PDIFF option; letter groups were determined using the PDMIX macro procedure. The relationship between carotenoid content in treatment diets and yolk color, carotenoid content, and carotenoid deposition efficiency was evaluated using Pearson correlation implemented in the CORR procedure. The threshold for statistical significance was defined as $p < 0.05$.

3. Results

3.1. Carotenoid Content in Experimental Diets

The experimental diets differed ($p < 0.001$) in the contents of all individual and total carotenoids (Table 2). The average contents of lutein and zeaxanthin in the experimental diets were similar (6.04 and 6.24 μg/g DM, respectively), although the treatments had a wide range of both carotenoids in the experimental diets. Treatment Bc 572 had the lowest dietary lutein content, while the content in the Riđan treatment was twice as high.

Opposite to lutein, the Bc 572 treatment had the highest dietary zeaxanthin content, which was twice as high as in the Riđan and Pajdaš treatments. Treatment Bc 572 also had the highest dietary contents of β-cryptoxanthin and β-carotene, resulting in the highest dietary content of total carotenoids among the treatments. Treatment Bc 572 had a 27% higher total carotenoid content than the treatment with the lowest content, Kekec.

Table 2. Average carotenoid content (µg/g DM) in treatment diets differing in corn hybrids.

Carotenoid	Dietary Treatment					SEM	*p*
	Bc 572	Kekec	Mejaš	Riđan	Pajdaš		
Lutein	3.72 [e]	5.64 [d]	6.53 [c]	7.36 [a]	6.94 [b]	0.09	<0.001
Zeaxanthin	9.99 [a]	5.87 [b]	5.78 [b]	4.57 [d]	4.98 [c]	0.10	<0.001
β-cryptoxanthin	1.67 [a]	0.83 [c]	1.02 [b]	0.75 [c]	0.73 [c]	0.05	<0.001
β-carotene	1.74 [a]	1.12 [c]	1.45 [b]	1.16 [c]	1.40 [b]	0.07	<0.001
Total carotenoids	17.13 [a]	13.45 [d]	14.78 [b]	13.86 [cd]	14.04 [c]	0.16	<0.001

Values in a row with different letters differ significantly (*p* < 0.05).

3.2. Yolk Color

The yolks from hens fed the tested dietary treatments differed (*p* < 0.001) in color determined by the YCF scale (Table 3), the difference being due to the differentiation of treatment Bc 572 from the other treatments. Despite the differences in the dietary carotenoid profile, the Kekec, Mejaš, Riđan and Pajdaš treatments yielded similar YCF scores. Of the CIE Lab color space parameters, only redness differed between the dietary treatments tested (*p* < 0.001; Table 2), and it followed the same relationship between treatments as the YCF scale. Dietary treatments tended to differ in yellowness (*p* = 0.065), with the lowest value detected in the treatment with the highest redness value, Bc 572.

Table 3. Average egg yolk color according to the Yolk Color Fan (YCF) scale and CIE Lab in eggs laid by hens fed dietary treatments differing in corn hybrids.

Carotenoid	Dietary Treatment					SEM	*p*
	Bc 572	Kekec	Mejaš	Riđan	Pajdaš		
YCF	10.8 [a]	9.92 [b]	9.96 [b]	9.98 [b]	9.83 [b]	0.08	<0.001
CIE Lab							
L*	67.27	67.64	67.76	67.70	67.35	0.21	0.351
a*	12.47 [a]	11.01 [b]	11.02 [b]	11.04 [b]	10.05 [b]	0.12	<0.001
b*	67.47	68.14	68.30	68.62	68.22	0.28	0.065

Values in a row with different letters differ significantly (*p* < 0.05).

The yolk YCF score and the redness and yellowness according to CIE Lab correlated with the content of individual and total carotenoids in the experimental diets. An increase in lutein content in the experimental diets was associated with a decrease in redness and YCF score (r = −0.73 and −0.76, respectively, *p* < 0.001) and an increase in yellowness (r = 0.43, *p* < 0.05). On the other hand, zeaxanthin, β-cryptoxanthin and β-carotene showed an opposite correlation compared to lutein. Zeaxanthin and β-cryptoxanthin correlated positively with redness and YCF (r = 0.77 and 0.78 for zeaxanthin and r = 0.70 and 0.71 for β-cryptoxanthin, respectively, *p* < 0.001) and negatively with yellowness (r = −0.42 and −0.46, respectively, *p* < 0.05). β-carotene and total carotenoids correlated with redness and YCF (r = 0.42 and 0.50 for β-carotene, *p* < 0.05, and r = 0.68 and 0.69 for total carotenoids, *p* < 0.001, respectively).

3.3. Yolk Carotenoid Profile

The content of individual and total carotenoids in yolks increased in all treatments during the first two weeks of the experiment (Figure 1). After the second week, the content of individual carotenoids showed weekly fluctuations, which were more pronounced for β-cryptoxanthin and β-carotene than for lutein and zeaxanthin. Weekly fluctuations in the total yolk carotenoid content were less evident compared to individual carotenoids. No significant decrease in the yolk content of the determined carotenoids was observed by the end of the experiment.

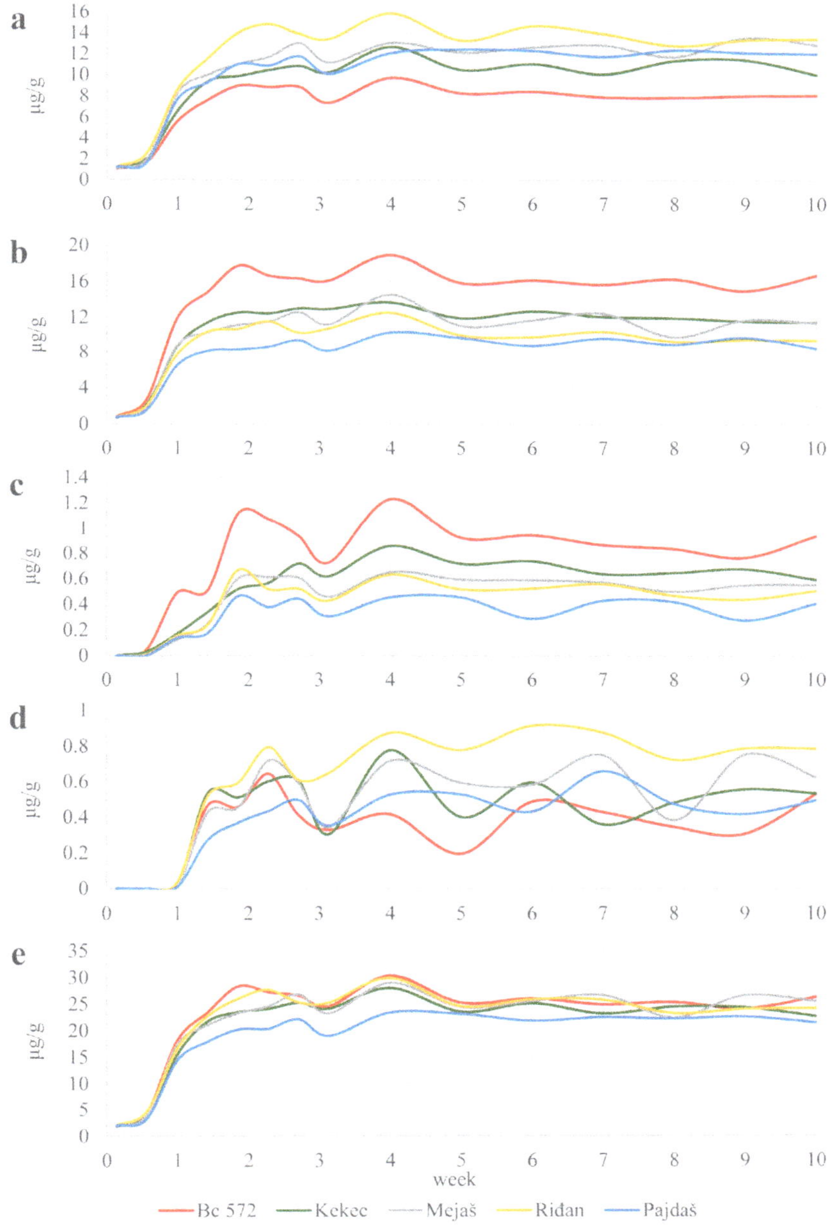

Figure 1. Changes in yolk contents of lutein (**a**), zeaxanthin (**b**), β-cryptoxanthin (**c**), β-carotene (**d**) and total carotenoids (**e**) in eggs laid by hens fed dietary treatments differing in corn hybrid.

The yolks of the experimental treatments differed in the content of individual and total carotenoids ($p < 0.001$; Table 4). The average content of lutein and zeaxanthin in the yolks was 11.51 and 12.00 µg/g, and their content was up to 25 times higher than the content of β-cryptoxanthin and β-carotene, on average, 0.61 and 0.56 µg/g, respectively. Treatment Bc 572 had the lowest lutein content and the highest content of zeaxanthin, β-cryptoxanthin and total carotenoids. Treatment Riđan had the highest contents of lutein and β-carotene. The contents of lutein, zeaxanthin and β-cryptoxanthin increased with their increase in the experimental diets ($r = 0.93$, 0.93 and 0.83, respectively, $p < 0.001$). On the other hand, the yolk content of β-carotene decreased with its content in the experimental diets ($r = -0.61$, $p < 0.001$).

Table 4. Average yolk carotenoid content (µg/g) in eggs laid by hens fed dietary treatments differing in corn hybrids.

Carotenoid	Dietary Treatment					SEM	p
	Bc 572	Kekec	Mejaš	Riđan	Pajdaš		
Lutein	8.31 [e]	10.92 [d]	12.52 [b]	13.91 [a]	11.88 [c]	0.19	<0.001
Zeaxanthin	16.36 [a]	12.35 [b]	11.78 [b]	10.29 [c]	9.22 [d]	0.22	<0.001
β-cryptoxanthin	0.92 [a]	0.69 [b]	0.56 [c]	0.51 [d]	0.39 [e]	0.02	<0.001
β-carotene	0.40 [d]	0.51 [c]	0.60 [b]	0.79 [a]	0.48 [cd]	0.03	<0.001
Total carotenoids	25.99 [a]	24.47 [b]	25.47 [ab]	25.49 [ab]	21.97 [c]	0.40	<0.001

Values in a row with different letters differ significantly ($p < 0.05$).

3.4. Carotenoid Deposition Efficiency in Yolks

The deposition efficiency of lutein and zeaxanthin (on average, 25.52 and 26.05%, respectively) in the yolks of the tested treatments was up to 10 times higher than the deposition efficiency of β-cryptoxanthin and β-carotene (on average, 8.30 and 5.65%, respectively). The tested treatments differed in the deposition efficiency of all individual and total carotenoids ($p < 0.001$; Table 5). Treatment Pajdaš achieved the lowest values of deposition efficiency for all individual carotenoids except β-carotene, and the deposition efficiency of total carotenoids was similar to the result of treatment Bc 572, despite the high total carotenoid content in the experimental diet of treatment Bc 572. In addition, treatments Kekec, Mejaš and Riđan resulted in a similar deposition efficiency of total carotenoids, regardless of the differences in the carotenoid profile of the experimental diets.

Table 5. Average carotenoid deposition efficiency (%) in yolks from eggs laid by hens fed dietary treatments differing in corn hybrids.

Carotenoid	Dietary Treatment					SEM	p
	Bc 572	Kekec	Mejaš	Riđan	Pajdaš		
Lutein	28.46 [a]	26.37 [b]	26.19 [b]	24.81 [b]	21.75 [d]	0.60	<0.001
Zeaxanthin	20.87 [d]	28.62 [ab]	27.76 [b]	29.53 [a]	23.49 [b]	0.60	<0.001
β-cryptoxanthin	7.00 [c]	11.21 [a]	7.52 [c]	8.87 [b]	6.90 [c]	0.32	<0.001
β-carotene	2.94 [d]	6.23 [b]	5.68 [b]	8.95 [a]	4.45 [c]	0.35	<0.001
Total carotenoids	19.31 [b]	24.74 [a]	23.50 [a]	24.15 [a]	19.85 [b]	0.50	<0.001

Values in a row with different letters differ significantly ($p < 0.05$).

The deposition efficiency of all individual and total carotenoids correlated negatively with their content in the experimental diets (lutein: $r = -0.62$, $p < 0.001$; zeaxanthin: $r = -0.68$, $p < 0.001$; β-cryptoxanthin: $r = -0.43$, $p < 0.05$; β-carotene: $r = -0.79$, $p < 0.001$; total carotenoids: $r = -0.63$, $p < 0.001$).

4. Discussion

The only source of carotenoids in the experimental diets was corn; therefore, lutein, zeaxanthin, β-cryptoxanthin and β-carotene, as the main carotenoids in corn [20], were found in the diets. Consequently, the differences between the experimental diets in the carotenoid profiles were attributed to the differences in the carotenoid profiles of the commercial corn hybrids used. Comparing the experimental diets in terms of their carotenoid profile with other studies is somewhat difficult because the current research focuses on biofortified corn. In those studies, commercial yellow corn has typically been used as the control to which biofortified corn has been compared. Experimental diets prepared with commercial yellow corn in these studies have lower levels of total carotenoids than the diets in the present study. Liu et al. [14] reported that commercial yellow corn contained 24.63 nmol/g, i.e., 13.89 µg/g of total carotenoids, and the resulting diet containing 600 g/kg of corn corresponds to a total carotenoid content of 8.33 µg/g. In addition, Moreno et al. [15] reported that the experimental diet containing 620.6 g/kg of corn had 9.2 µg/g of total carotenoids, while Ortiz et al. [22] reported 5.7 µg/g of total carotenoids in a diet containing 565 g/kg of commercial yellow corn.

Although studies have shown that biofortification generally results in increased carotenoid content in the experimental hen diets, this increase could represent an increase in the content of some specific carotenoids, which, consequently, does not result in an increase in total carotenoids. The experimental diets in the present study had higher contents of lutein, zeaxanthin, β-cryptoxanthin and β-carotene but did not contain violaxanthin and astaxanthin, which were present in the diets with corn fortified with ketocarotenoids in the study by Moreno et al. [15], resulting in similar total carotenoid contents (13.81 µg/g vs. 14.65 µg/g on average in the present study). Similarly, high β-cryptoxanthin corn in the study by Liu et al. [14] provided a similar amount of β-cryptoxanthin and β-carotene (4.71 and 5.31 nmol/g for corn varieties, or 1.56 and 1.71 µg/g, calculated on the basis of corn proportion in the diet, respectively) as treatment Bc 572 in the present study but with a lower total carotenoid amount (13.89 µg/g, calculated on the basis of corn proportion in the diet, vs. 17.13 µg/g, respectively). Therefore, some existing commercial corn hybrids have comparable levels of individual or total carotenoids to carotenoid-biofortified corn, i.e., they could provide their similar levels to the hen through the diet. This implies that some high-yielding commercial corn hybrids, depending on yolk pigmenting ability, could be used in targeted production as corn for laying hens.

The next step in evaluating the suitability of commercial corn hybrids as a single pigment in hen diets was to determine their yolk pigmentation ability. The results of the YCF values obtained show that treatment Bc 572 was the only commercial hybrid able to pigment the yolk to the color acceptable in most markets, while the other treatments pigment the yolk in such way that the color is acceptable in markets where a paler yolk color is desired. Similar values to treatments tested in the present study were achieved by the diet containing 620.6 g/kg of genetically engineered corn fortified with carotenoids in the study by Moreno et al. [15], while the diet containing 565 g/kg of Orange Corn, non-GMO corn biofortified to increase the total and provitamin A carotenoid content in the study by Ortiz et al. [22], achieved even lower values (9 ± 0.3). Diets prepared with the yellow commercial corn from these two studies (YCF scores ~ 6 and 4, respectively) achieved a color significantly lower than the YCF scores of the present study. When considering the CIE Lab, the increase in yolk YCF scores is accompanied by the increase in yolk redness (r = 0.96, $p < 0.001$), suggesting that carotenoids affecting redness contribute most to the changes in the YCF scale. The dietary treatments in the present study resulted in eggs that varied in yolk redness from 10.23 to 12.47, which is higher than the values for biofortified corn in studies by Liu et al. [14] and Ortiz et al. [22] (~5 and 4, respectively).

The commercial corn hybrids in the present study achieved yolk colors comparable to various pigments in other studies. Skřivan et al. [23] reported a similar YCF score (10.55) and redness values (11.51) of yolks from hens fed diets based on 350 g/kg of corn and 280 g/kg wheat supplemented with 950 µg/g of marigold flower extract, while treatment Bc 572 in the present study achieved similar yolk YCF score (10.55) as hens fed diets containing 600 g/kg of wheat supplemented with 15 µg/g of spirulina in a study by Zahroojian et al. [24]. However, the yolk color intensity in the present study was not as high as that of the commercial synthetic pigment containing canthaxanthin added to the standard diet at a concentration of 8 mg/kg (YCF 13.47, a* 18.76) [10]. The higher pigmentation effect of canthaxanthin compared to the carotenoids present in corn is due to its higher deposition efficiency in the yolk; 37 to 50% of the ingested canthaxanthin is deposited in the yolk [2]. However, although the dietary treatments were not as successful as the synthetic pigments in achieving high YCF scores, the eggs from the present study provide more carotenoids in the human diet, especially lutein and zeaxanthin, at no additional cost to the pigment source. On the other hand, the YCF scores obtained were close to the value of 10.7 obtained in the study by Englmaierová et al. [25], who used a combination of synthetic carotenoids with 2 mg/kg of canthaxanthin and 1.5 mg/kg of ethyl ester of β-apo-8′-carotenoic acid in hen diet.

Based on the correlations between the dietary content of individual carotenoids and color scores, dietary lutein contributed to yolk yellowness, which decreased the YCF score. On the other hand, zeaxanthin, β-cryptoxanthin and β-carotene contributed to yolk redness, which increased the YCF score. These correlations suggest that an adequate profile of commercial corn hybrids for desirable yolk pigmentation has elevated levels of the latter three carotenoids, as found in treatment Bc 572. This observation is in agreement with Liu et al. [14] and Ortiz et al. [22], who also showed that increased zeaxanthin and β-cryptoxanthin in diets as a result of increased content in corn grain resulted in higher yolk color intensity.

The accumulation of carotenoids in the yolks occurred after the fourth day, after the experimental diets were offered to the hens, and saturation was reached after the second week. The duration of the accumulation phase in the present study falls within the range reported in studies with biofortified corn. Moreno et al. [15] reported that saturation was reached after the third week in all dietary treatments, while Ortiz et al. [22] showed that maximum carotenoid accumulation was reached on the 12th day of feeding with dietary treatments. During the experimental period, the yolk levels of lutein and zeaxanthin were higher and more constant than those of β-cryptoxanthin and β-carotene. Moreover, the weekly fluctuations of β-carotene in the yolk showed a wide range in all the treatments tested. These fluctuations most likely reflect the utilization of β-cryptoxanthin and β-carotene by the hen; as previously reported [26], they are converted to vitamin A, with β-carotene being more readily converted than β-cryptoxanthin [14]. As yolk carotenoid levels remained constant until the end of the experiment, commercial corn hybrids provided a stable source of carotenoids for yolk pigmentation.

In agreement with previous reports [8,14], yolk carotenoid profiles of the tested treatments reflected the carotenoid profile of the experimental diets, with zeaxanthin dominating in the eggs of treatments Bc 572 and Kekec and lutein in the eggs of treatments Mejaš, Riđan and Pajdaš. The high correlation coefficients between the dietary and yolk contents of lutein, zeaxanthin and β-cryptoxanthin confirm these considerations. On the other hand, the β-carotene content decreased with the dietary content, in accordance with the preferential conversion to vitamin A. The experimental diets had the same content of ingredients and the other ingredients except corn were from the same batch, thus unifying the effect of the diet matrix on carotenoid bioavailability. From compounds present in the experimental diets, oleic acid was the most important for carotenoids; corn contains 26.8% while sunflower oil contains 20% of this fatty acid [27]. Monounsaturated fatty acids improve the absorption of polar carotenoids, especially when associated with high ME density (11.6 MJ/kg) [28].

The yolk total xanthophyll content reached values in the range of 21.5 to 25.5 μg/g, which was close to lower values in yolks from ecological eggs (20.5–33.5 μg/g) but close to higher values for yolks from free range (12.6–25.5 μg/g), barn (12.1–26.8 μg/g) or eggs (7.3–29.3 μg/g) purchased in local German supermarkets [29]. This comparison suggests that commercial corn hybrids provide more carotenoids than most free-range, barn and cage eggs, in which canthaxanthin, β-apo-8′-ethyl ester and citranaxanthin were found in considerable amounts. Moreover, the yolks from hens fed corn hybrids in the present study provided more xanthophylls than conventional cage, cage free, cage-free and free-range/pasture and free-range/pasture-organic eggs (21.5, 22.1, 18.2, 18.8, and 22.6 µg/g, respectively) purchased from the local market in the study by Ortiz et al. [22]. The yolks from the later study had YCF scores between 6 and 8, implying that commercial corn hybrids could result in higher color intensity with only a small increase in xanthophyll content in the yolks.

Compared to biofortified corn, commercial corn hybrids were within the range of total carotenoid contents in yolk reported in previous studies. While Liu et al. [14] reported lower values for high-β-carotene and high-β-cryptoxanthin corn (23.61 and 25.86 nmol/g, i.e., 13.37 and 14.61 µg/g, respectively), Ortiz et al. [22] reported a higher value for Orange Corn (29.4 µg/g). Moreno et al. [15] reported results in the freeze-dried yolks and based on the fact that the yolk contains approximately 50% moisture [30], corn enriched in carotenoids resulted in similar contents of total carotenoids (57.5 µg/g of freeze-dried yolk), while corn enriched in ketocarotenoids resulted in twice lower content (26.18 µg/g of freeze-dried yolk). In these previous studies, the biofortification of corn in the diet of hens resulted in increased zeaxanthin and β-cryptoxanthin content and color intensity in the yolks, and a similar result was found with treatment Bc 572 in the present study.

The deposition efficiencies of lutein (on average 25.52%) and zeaxanthin (on average 26.05%) were higher than those of β-cryptoxanthin (on average 8.30%) and β-carotene (on average 5.65%). In addition to the preferential utilization of β-cryptoxanthin and β-carotene as vitamin A, these results also reflect the easier absorption of the more polar carotenoids lutein and zeaxanthin [31]. Furthermore, van het Hof [32] has shown that lutein is five times more bioavailable than β-carotene in humans. The deposition efficiencies of individual carotenoids were within the range of values reported in previous studies [8,33,34]. These studies reported variable deposition efficiencies but also differed in the carotenoid source and its inclusion level in the diet. For example, Karadas et al. [8] reported deposition efficiencies of zeaxanthin ranging from 9.1–30.3% and β-carotene ranging from 0.4–3.8% for diets containing 20 g/kg of lucerne extract, marigold extract and tomato powder. Hammershøj et al. [29] reported values in the range of 18.8–27.4 for lutein, 18.9–27.9 for zeaxanthin and 0.3–1.0 for β-carotene in diets supplemented with 70 g/day/hen of different carrot varieties. Thus, the values reported in the present study show good deposition efficiency of corn carotenoids, including β-carotene, into the egg yolk.

In addition to the differences between individual carotenoids, the deposition efficiency of corn carotenoids also varied with their concentration—the levels of all individual and total carotenoids decreased with increasing dietary content. This relationship between the deposition efficiency into yolk and dietary content has been reported previously for lutein [8,34]. It seems that other corn carotenoids follow the same relationship, in agreement with their decreasing bioaccessibility from the matrix with increasing content in the grain [35]. However, the tested corn hybrids as the only source of carotenoids in hen diets resulted in high yolk carotenoid content and color intensity for the majority of world markets, despite the observed relationship between the deposition efficiency and the dietary content of carotenoids.

5. Conclusions

Commercial high-yielding corn hybrids may differ in carotenoid content, but the present study showed that those with high carotenoid contents could be the only source of pigments for yolk pigmentation in hen diets. More so, the dietary treatments resulted in the high yolk content of lutein and zeaxanthin, comparable to biofortified corn, which has been widely studied recently, with no decrease in color intensity compared to synthetic pigment sources widely used in poultry production. Corn hybrids containing higher levels of zeaxanthin and β-cryptoxanthin even resulted in higher yolk color intensity (higher YCF score and redness values). The results of the study suggest that more attention should be given to the selection of existing commercial corn hybrids for poultry; the inclusion of these hybrids in the diet could lead to a reduction in egg production costs.

Author Contributions: Conceptualization, K.K. and D.G.; methodology, K.K., M.D. and Z.J.; validation, K.K., D.G. and Z.J.; formal analysis, K.K., M.D., G.K. and D.B.; investigation, K.K., M.D., G.K., D.B. and Z.J.; resources, D.G. and Z.J.; data curation, K.K.; writing—original draft preparation, K.K.; writing—review and editing, D.G., M.D., G.K., D.B. and Z.J.; visualization, K.K.; supervision, D.G.; project administration, K.K. and D.G.; funding acquisition, D.G. All authors have read and agreed to the published version of the manuscript.

Funding: This research received no external funding.

Institutional Review Board Statement: The study was conducted according to the European guidelines for the care and use of animals used for scientific purposes and approved by the Bioethics Committee for the protection and welfare of animals at the University of Zagreb Faculty of Agriculture (KLASA 114-04/17-03/02, URBROJ 251-71-01-17-1; date of approval 6 June 2017).

Informed Consent Statement: Not applicable.

Data Availability Statement: The data presented in this study are available on request from the corresponding author.

Conflicts of Interest: The authors declare no conflict of interest. The funders had no role in the design of the study; in the collection, analyses, or interpretation of data; in the writing of the manuscript, or in the decision to publish the results.

References

1. Viola, I.; Marinelli, A. Life cycle assessment and environmental sustainability in the food system. *Agric. Agric. Sci. Procedia* **2016**, *8*, 317–323. [CrossRef]
2. Grashorn, M. Feed additives for influencing chicken meat and egg yolk color. In *Handbook on Natural Pigments in Food and Beverages*; Carle, R., Schweiggert, R., Eds.; Woodhead Publishing: Cambridge, UK, 2016; pp. 283–302.
3. Abdel-Aal, E.S.M.; Akhtar, H.; Zaheer, K.; Ali, R. Dietary sources of lutein and zeaxanthin carotenoids and their role in eye health. *Nutrients* **2013**, *5*, 1169–1185. [CrossRef]
4. Huang, Y.M.; Dou, H.L.; Huang, F.F.; Xu, X.R.; Zou, Z.Y.; Lu, X.R.; Lin, X.M. Changes following supplementation with lutein and zeaxanthin in retinal function in eyes with early age-related macular degeneration: A randomised, double-blind, placebo-controlled trial. *Br. J. Ophthalmol.* **2015**, *99*, 371–375. [CrossRef] [PubMed]
5. BCC Research. The Global Market for Carotenoids. 2015. Available online: https://www.bccresearch.com/market-research/food-and-beverage/the-global-market-for-carotenoids.html (accessed on 23 August 2021).
6. Karunajeewa, H.; Hughes, R.J.; McDonald, M.W.; Shenstone, F.S. A review of factors influencing pigmentation of egg yolks. *Worlds Poult. Sci. J.* **1984**, *40*, 52–65. [CrossRef]
7. Beardsworth, P.M.; Hernandez, J.M. Yolk colour—An important egg quality attribute. *Int. Poult. Prod.* **2004**, *12*, 17–18.
8. Karadas, F.; Grammenidis, E.; Surai, P.F.; Acamovic, T.; Sparks, N.H.C. Effects of carotenoids from lucerne, marigold and tomato on egg yolk pigmentation and carotenoid composition. *Br. Poult. Sci.* **2006**, *47*, 561–566. [CrossRef] [PubMed]
9. van Ruth, S.M.; Koot, A.H.; Brouwer, S.E.; Boivin, N.; Carcea, M.; Zerva, C.N.; Haugen, J.-E.; Höhl, A.; Köroglu, D.; Mafra, I.; et al. Eggspectation: Organic egg authentication method challenged with produce from ten different countries. *Qual. Assur. Saf. Crop. Foods* **2013**, *5*, 7–14. [CrossRef]
10. Kljak, K.; Carović-Stanko, K.; Kos, I.; Janječić, Z.; Kiš, G.; Duvnjak, M.; Safner, T.; Bedeković, D. Plant carotenoids as pigment sources in laying hen diets: Effect on yolk color, carotenoid content, oxidative stability and sensory properties of eggs. *Foods* **2021**, *10*, 721. [CrossRef] [PubMed]
11. Skřivan, M.; Englmaierová, M.; Skřivanová, E.; Bubancová, I. Increase in lutein and zeaxanthin content in the eggs of hens fed marigold flower extract. *Czech J. Anim. Sci.* **2015**, *60*, 89–96. [CrossRef]
12. Mellado-Ortega, E.; Hornero-Méndez, D. Carotenoids in cereals: An ancient resource with present and future applications. *Phytochem. Rev.* **2015**, *14*, 873–890. [CrossRef]
13. Zhu, C.; Farré, G.; Zanga, D.; Lloveras, J.; Michelena, A.; Ferrio, J.P.; Voltas, J.; Slafer, G.; Savin, R.; Albajes, R.; et al. High-carotenoid maize: Development of plant biotechnology prototypes for human and animal health and nutrition. *Phytochem. Rev.* **2018**, *17*, 195–209. [CrossRef]
14. Liu, Y.Q.; Davis, C.R.; Schmaelzle, S.T.; Rocheford, T.; Cook, M.E.; Tanumihardjo, S.A. β-Cryptoxanthin biofortified maize (*Zea mays*) increases β-cryptoxanthin concentration and enhances the color of chicken egg yolk. *Poult. Sci.* **2012**, *91*, 432–438. [CrossRef]
15. Moreno, J.A.; Díaz-Gómez, J.; Fuentes-Font, L.; Angulo, E.; Gosálvez, L.F.; Sandmann, G.; Portero-Otin, M.; Capell, T.; Zhu, C.; Christou, P.; et al. Poultry diets containing (keto) carotenoid-enriched maize improve egg yolk color and maintain quality. *Anim. Feed Sci. Technol.* **2020**, *260*, 114334. [CrossRef]
16. Zanga, D.; Capell, T.; Slafer, G.A.; Christou, P.; Savin, R. A carotenogenic mini-pathway introduced into white corn does not affect development or agronomic performance. *Sci. Rep.* **2016**, *6*, 38288. [CrossRef] [PubMed]

17. National Research Council (NRC). *Nutrient Requirements of Poultry*, 9th ed.; National Academy Press: Washington, DC, USA, 1994.
18. Báblona TETRA Ltd. Báblona TETRA Commercial Layer Management Guide. Available online: http://www.babolnatetra.com/wp-content/uploads/2019/07/tetra-sl-ps.pdf (accessed on 15 March 2017).
19. International Organization for Standardization (ISO). *Animal Feeding Stuffs—Determination of Moisture and Other Volatile Matter Content*; No. 6494; International Organization for Standardization: Geneva, Switzerland, 1999.
20. Kurilich, A.C.; Juvik, J.A. Quantification of carotenoid and tocopherol antioxidants in *Zea mays*. *J. Agric. Food Chem.* **1999**, *47*, 1948–1955. [CrossRef]
21. Surai, P.F.; Speake, B.K.; Wood, N.A.; Blount, J.D.; Bortolotti, G.R.; Sparks, N.H. Carotenoid discrimination by the avian embryo: A lesson from wild birds. *Comp. Biochem. Physiol. B Biochem. Mol. Biol.* **2001**, *128*, 743–750. [CrossRef]
22. Ortiz, D.; Lawson, T.; Jarrett, R.; Ring, A.; Scoles, K.L.; Hoverman, L.; Rocheford, E.; Karcher, D.M.; Rocheford, T. Biofortified orange corn increases xanthophyll density and yolk pigmentation in egg yolks from laying hens. *Poult. Sci.* **2021**, *100*, 101117. [CrossRef]
23. Skřivan, M.; Marounek, M.; Englmaierova, M.; Skřivanová, E. Effect of increasing doses of marigold (*Tagetes erecta*) flower extract on eggs carotenoids content, colour and oxidative stability. *J. Anim. Feed Sci.* **2016**, *25*, 58–64. [CrossRef]
24. Zahroojian, N.; Moravej, H.; Shivazad, M. Comparison of marine algae (*Spirulina platensis*) and synthetic pigment in enhancing egg yolk colour of laying hens. *Br. Poult. Sci.* **2011**, *52*, 584–588. [CrossRef]
25. Englmaierová, M.; Skrivan, M.; Bubancová, I.A. Comparison of lutein, spray-dried *Chlorella*, and synthetic carotenoids effects on yolk colour, oxidative stability, and reproductive performance of laying hens. *Czech J. Anim. Sci.* **2013**, *58*, 412–419. [CrossRef]
26. Hencken, H. Chemical and physiological behavior of feed carotenoids and their effects on pigmentation. *Poult. Sci.* **1992**, *71*, 711–717. [CrossRef]
27. Sauvant, D.; Perez, J.M.; Tran, G. *Tables of Composition and Nutritive Value of Feed Materials*; INRA Editions: Versailles, France, 2004.
28. Papadopoulos, G.A.; Chalvatzi, S.; Kopecký, J.; Arsenos, G.; Fortomaris, P.D. Effects of dietary fat source on lutein, zeaxanthin and total carotenoids content of the egg yolk in laying hens during the early laying period. *Br. Poult. Sci.* **2019**, *60*, 431–438. [CrossRef]
29. Schlatterer, J.; Breithaupt, D.E. Xanthophylls in commercial egg yolks: Quantification and identification by HPLC and LC-(APCI) MS using a C30 phase. *J. Agric. Food Chem.* **2006**, *54*, 2267–2273. [CrossRef] [PubMed]
30. Heying, E.K.; Tanumihardjo, J.P.; Vasic, V.; Cook, M.; Palacios-Rojas, N.; Tanumihardjo, S.A. Biofortified orange maize enhances β-cryptoxanthin concentrations in egg yolks of laying hens better than tangerine peel fortificant. *J. Agric. Food Chem.* **2014**, *62*, 11892–11900. [CrossRef] [PubMed]
31. Priyadarshani, A.M.B. A review on factors influencing bioaccessibility and bioefficacy of carotenoids. *Crit. Rev. Food Sci. Nutr.* **2017**, *57*, 1710–1717. [CrossRef] [PubMed]
32. van het Hof, K.H.; Brouwer, I.A.; West, C.V.; Haddeman, E.; Steegers-Theunissen, R.P.M.; van Dusseldorp, M.; Weststrate, J.A.; Eskes, T.K.; Hautvast, J.G. Bioavailability of lutein from vegetables is 5 times higher than that of β-carotene. *Am. J. Clin. Nutr.* **1999**, *70*, 261–268. [CrossRef]
33. Hammershøj, M.; Kidmose, U.; Steenfeldt, S. Deposition of carotenoids in egg yolk by short-term supplement of coloured carrot (*Daucus carota*) varieties as forage material for egg-laying hens. *J. Sci. Food Agric.* **2010**, *90*, 1163–1171. [CrossRef]
34. Leeson, S.; Caston, L. Enrichment of eggs with lutein. *Poult. Sci.* **2004**, *83*, 1709–1712. [CrossRef] [PubMed]
35. Zurak, D.; Grbeša, D.; Duvnjak, M.; Kiš, G.; Međimurec, T.; Kljak, K. Carotenoid content and bioaccessibility in commercial maize hybrids. *Agriculture* **2021**, *11*, 586. [CrossRef]

sustainability

MDPI

Review

Honeybee and Plant Products as Natural Antimicrobials in Enhancement of Poultry Health and Production

Erinda Lika [1], Marija Kostić [2], Sunčica Vještica [3], Ivan Milojević [4] and Nikola Puvača [5,*]

[1] Faculty of Veterinary Medicine, Agricultural University of Tirana, Koder Kamez, 1029 Tirana, Albania; elika@ubt.edu.al

[2] Faculty of Hotel Management and Tourism Vrnjačka Banja, University of Kragujevac, Vojvođanska Ulica bb, 36210 Vrnjačka Banja, Serbia; marija.kostic@kg.ac.rs

[3] The Faculty of Applied Ecology, University Metropolitan Belgrade, Požeška 83a, 11000 Belgrade, Serbia; suncica.vjestica@futura.edu.rs

[4] Ministry of Defense, The Human Resource Sector, Nemanjina 15, 11000 Belgrade, Serbia; drimilojevic@gmail.com

[5] Department of Engineering Management in Biotechnology, Faculty of Economics and Engineering Management in Novi Sad, University Business Academy in Novi Sad, Cvećarska 2, 21000 Novi Sad, Serbia

* Correspondence: nikola.puvaca@fimek.edu.rs; Tel.: +381-65-219-1284

Abstract: The quality and safety attributes of poultry products have attracted increasing widespread attention and interest from scholarly groups and the general population. As natural and safe alternatives to synthetic and artificial chemical drugs (e.g., antibiotics), botanical products are recently being used in poultry farms more than 60% of the time for producing organic products. Medicinal plants, and honeybee products, are natural substances, and they were added to poultry diets in a small amount (between 1% and 3%) as a source of nutrition and to provide health benefits for poultry. In addition, they have several biological functions in the poultry body and may help to enhance their welfare. These supplements can increase the bodyweight of broilers and the egg production of laying hens by approximately 7% and 10% and enhance meat and egg quality by more than 25%. Moreover, they can improve rooster semen quality by an average of 20%. Previous research on the main biological activities performed by biotics has shown that most research only concentrated on the notion of using botanical products as growth promoters, anti-inflammatory, and antibacterial agents. In the current review, the critical effects and functions of bee products and botanicals are explored as natural and safe alternative feed additives in poultry production, such as antioxidants, sexual-stimulants, immuno-stimulants, and for producing healthy products.

Keywords: honeybee; medicinal plants; quality; antimicrobials; poultry; health

Citation: Lika, E.; Kostić, M.; Vještica, S.; Milojević, I.; Puvača, N. Honeybee and Plant Products as Natural Antimicrobials in Enhancement of Poultry Health and Production. *Sustainability* **2021**, *13*, 8467. https://doi.org/10.3390/su13158467

Academic Editor: George K. Symeon

Received: 16 June 2021
Accepted: 27 July 2021
Published: 29 July 2021

Publisher's Note: MDPI stays neutral with regard to jurisdictional claims in published maps and institutional affiliations.

1. Introduction

The poultry industry has recently faced many challenges, including economic recession, climate change, disease, and overuse of antibiotics. Enhancing animal welfare, production, and health is a significant request for all poultry farms and provides safe and organic products to consumers [1]. Organic production refers to the final product quality and extends to the whole production process under high quality and security control [2,3]. Therefore, today's global trend is to lessen the usage of synthetic prophylactic and therapeutic drugs (such as antibiotics) in poultry farms and find safe and healthy natural replacements [4,5]. The prospect of using new natural materials as dietary supplements in poultry diets instead of antibiotics has been recently studied [6]. These materials have to allow the production of safe and high-quality food. For decades, antibiotics were commonly used in poultry farms to preserve the gut ecosystem's equilibrium and improve chicken growth [7]. The overuse of these antibiotics in poultry farms, as growth promoters for improving feed conversion ratio and growth, has resulted in several adverse effects such as the development of antimicrobial resistance (AMR) and transference, and the residues remained in consumed meat [8–11]. Several strategies have been applied to deal with this global trend. One of these strategies is to use botanicals as natural substances in poultry production to improve poultry welfare and produce organic meat and eggs [12]. The botanical products can be defined as a natural substance derived from natural products such as plant extracts, fruits, and bee products, added to the animal diet to provide medical or health benefits, including preventing or treating a disease [13–15]. Botanicals appear to have a wide variety of uses in the nutrition and production of poultry [16,17]. In various ways, they are distinct from other dietary supplements, such as their ability to play positive biological roles in an animal's body and their capability to improve its health status without leaving any traces in consumed meat [18–21]. The probable mechanism of action of plant products has been found to be through their beneficial effect on gut microflora by reducing the number of pathogenic organisms, resulting in increased nutrient availability for the host [22–25]. Due to their nutritional and medical properties, herbal and bee products are now used as dietary supplements in poultry production on a large scale [26–32]. Additionally, they are progressively used for in vivo feeding techniques as multiple functions such as nutrients for growth and immune stimulation in young chicks [33]. Because the significant benefits in poultry production of both products originated from bees and plants, we decided to unify those products in this review. Furthermore, those products express their efficacy in terms of meat performance, carcass traits, meat quality, immunity, egg quality, and blood parameters, along with the costs and returns on investment to establish the usefulness as an alternative to antibiotics growth promoters, improving their importance.

2. Biological Functions of Botanicals

Botanicals, in general, are natural chemical components with a significant part in the modification and maintenance of normal physiological activities that promote the health of the host [34–36]. They can be divided into botanicals that are still under investigation, such as plant essential oils, and those whose mode of action is supported scientifically [37]. Due to their considerable growth and health benefits, botanicals are widely used in poultry diets as commercial additives and alternative feed supplements to improve animal production and welfare [38]. A healthy gut is generally the cornerstone of optimal poultry growth performance [39]. Sugiharto [40] found that gut health could be affected by various factors. When gut health and function are damaged, the digestion and absorption of nutrients are impacted, which impacts the growth and productivity of poultry. On the other hand, when they are enhanced, the growth and productivity of poultry are improved. Sugiharto [40] also concluded that botanicals could improve the gut ecosystem and the immune functions of poultry and result in growth promotion and prevention or treatment of enteric infections. Botanicals can positively affect the balance of intestinal microbiota, which play a vital role in regulating metabolism, intestinal epithelial proliferation, and vitamin synthesis [41–44]. Many botanicals, in the form of prebiotics, probiotics, and symbiotics, are used to promote poultry gut health [45].

Additionally, these natural products could help in protecting the host against infectious diseases [46]. Plant essential oils have been used routinely in poultry farms for keeping poultry healthy and enhancing their productive performance [47]. Abo Ghanima et al. [47] showed significant and positive results on layers performance, egg quality, hematological traits, blood chemistry, and immunity with the dietary addition of rosemary and cinnamon essential oils. The experiment was conducted on a total of 5000 ISA brown laying hens during the production stages from 28 to 76 weeks of age; the application of 300 mg/kg of essential oils showed promising results. These essential oils contain active substances that have a beneficial impact on physiological processes and have therapeutic properties such as anti-inflammatory and antibacterial effects [48]. In general, previous literature focusing on the biological activity of botanicals has shown that most studies only discuss the concept of their use as growth promoters and anti-inflammatory and antibacterial agents. Our review discusses other critical biological functions using recent natural products as safe alternative feed additives in poultry production and health. The recent findings and applications concerning the bee (Table 1) and botanical (Table 2) products as natural supplementations adding in poultry diets for improving poultry productivity and enhancing their immunity are summarized, including their mechanisms of action.

Table 1. Effects of dietary addition of bee products in daily poultry production.

Additive	Poultry Species	Additive Concentrations	Obtained Results	Source
Honey	Japanese quails	22 g/L	Improved weight gain, feed intakes, and lower feed conversion ratio; improved immune system, and blood parameters; improved meat quality	[49–51]
Royal jelly	Laying hens	100–400 mg/kg	Increased egg production, improved welfare, and improved immunity; improved performance, egg quality, serum biochemistry and intestinal morphology	[52,53]
Bee venom	Broiler chickens	10–500 µg/kg	Improved production results, fatty acid composition, and antioxidant capacity; Better early development of chickens' digestive system and a helpful tool against short bowel syndrome	[54,55]
Bee pollen	Laying hens and quails	500–1500 mg/kg	Improved production results and biochemical blood parameters with the dietary addition of 500 mg/kg; improved egg production performance, blood biochemical and immunological response	[56,57]
Bee propolis	Broiler chickens	0.5–1.5%	Enhanced immunity of chickens with dietary addition in the concentration of 1.5%; improved performance, carcass traits and blood parameters	[58,59]
	Broiler chickens	200–400 mg/kg	The dietary concentration of 200 to 400 mg/kg improved blood lipid status	[60]
	Laying hens	250–1000 mg/kg	Increased body weight, egg production, and stimulate immunity; improved hematological and immunological parameters; improved performance, digestibility, egg production and egg quality under different environmental temperatures	[61–63]
	Japanese quails	1000 mg/kg	Retaining the performance and egg production at expedient levels under heat stress conditions; improved productivity of laying Japanese quails	[64,65]

Table 2. Effects of dietary addition of botanicals in daily poultry production.

Additive	Poultry Species	Additive Concentrations	Obtained Results	Source
Fenugreek seeds	Broiler chickens	1–3%	Improved production of chickens with the dietary addition of 1% to 3% of fenugreek seed powder	[66,67]

Table 2. *Cont.*

Additive	Poultry Species	Additive Concentrations	Obtained Results	Source
Black cumin	Broiler chickens	5–10 g/kg	Improved broiler performance and meat quality by enhancing antioxidant activities and suppressing lipid peroxidation in meat	[68,69]
Ginger	Broilers chickens	2–6 g/kg	Improved oxidative status; improved anticoccidial effects against experimentally induced coccidiosis; improved growth performance, haematological profiles, slaughter traits and gut morphometry	[70–72]
	Laying hens	0.25–0.75%	Improved antioxidant status and performance, decreased egg yolk cholesterol levels	[73]
Turmeric	Laying hens	0.5–3%	Improved performance, serum and egg yolk antioxidant status; improved blood metabolites and production performance characteristics; improved performance, egg quality traits, and blood parameters	[74–77]
	Broiler chickens	0.5–1 g/kg	Under heat stress, dietary addition of 0.5% increased the serum concentration of T3 and T4; improved blood constituents and performance	[78–80]
Thyme	Broiler chickens	1–3% 100–200 mg/kg	Possibility to replace antibiotics in the diet of chickens; Reduction of fat in chicken carcasses; improved growth, lipid oxidation, meat fatty acid composition and serum lipoproteins; improved growth performance microbial counts in the crop, small intestine and caecum	[81–83]
	Laying hens	300 mg/kg	Maintain production results, decreased serum glucose level, decrease plasma MDA during the heat stress; improved productive performance, egg quality traits, and blood parameters under cold stress conditions	[84–86]
	Japanese quails	0.01–1%	Increased antibody titer against sheep red blood cells (SRBC), lymphocyte proliferation, and respiratory bursting ability and decreased the delayed type of hypersensitivity; improved performance, some blood parameters and ileal microflora	[87,88]

Bee and botanical products express various protective effects when used as a dietary supplement in daily nutrition. Flavonoids, a major group of bee ingredients, are known as potential natural iron chelators and antioxidants. They appeared to play an important role as cardioprotective agents in doxorubicin-induced cardiac toxicity caused by the production of oxygen free radicals. The specific mechanism of action induced by flavonoids found in bee products is not yet fully elucidated. Some findings state that the cardio-protective potential of bee propolis could be associated with the radical scavenging action of caffeic acid phenethyl ester [89]. Dietary supplementation with propolis significantly reduces the serum concentrations of triglyceride, cholesterol, nitric oxide, very low-density lipoprotein (LDL), cholesterol, nitric oxide synthase (NOS), plasma glucose and malonaldehyde in fasting murine models, which improves the circulatory levels of high-density lipoprotein (HDL), cholesterol, and superoxide dismutase activity, indicating that propolis could play a role in normalizing the metabolism of circulatory fats [90]. According to the published data, the components of propolis can have potentially beneficial effects against microbial infections, hypertension, cardiovascular disorders, diabetes, and several chronic diseases. It is confirmed that several therapeutic impacts of bee products are due to their anti-oxidative, immunomodulatory, and anti-inflammatory activities. Some research lies behind the molecular mechanism of bee products' mediated protective effects [91]. Dysfunction of mitochondria is the major factor of ROS production, which leads to hypertension, diabetes, allergies, and asthma. The high production of ROS involves oxidative injury to cellular components (nucleic acids, proteins, and lipids) of tissues. Likewise, several other mechanisms are also involved in the generation of ROS. Enzymatic and non-enzymatic oxidation of polyunsaturated fatty acids such as arachidonic fatty acid. The interaction of unsaturated fatty acids with the trace metal ion $Fe3+$ resulted in the per-oxidation of fatty acids and the intensification of ROS production. ROS are also generated by NADPH oxidase. The components of bee propolis like quercetin, p-coumaric, caffeic, ferulic acids, caffeic acid phenethyl ester, and chrysin block the production of inflammatory mediators by suppressing the expression of lipoxygenases, cyclooxygenases, phospholipase A2, and nitric oxidase synthase [92]. Additionally, other bee components also inhibited the production of NF-kB by blocking its translocation to the nucleus [93].

Most of the researchers attributed the better performance of the poultry that were fed natural products to an improvement in palatability and a quick digestive effect. They further postulated that because of these natural products, the digestive tract would have been emptied earlier and feed consumption promoted. Ginger has been found to increase the secretion of gastrointestinal enzymes, including lipase, disaccharidase, and maltase [70]. Generally, improved performance may be attributed to the digestive enzymes such as protease and lipase, which are present as part of the plant's natural protective mechanisms. Reports have shown that natural products enhance animals' nutrient digestion and absorption because of their positive effect on gastric secretion, enterokinase, and digestive enzyme activities [94]. Even the dietary addition of natural products to laying hens expressed positive effects in better egg production and mass; the exact mechanism through which egg-laying performance is enhanced is not known. The higher performance of the laying hens may be due to antioxidant, antimicrobial, increased blood circulation, secretion of digestive enzymes, and reduction in feed oxidation. Despite increased levels of antioxidant enzymes and a reduction in MDA concentrations, the mechanism involved is not yet known [74]. Literature has shown that polyphenolic flavonoids in plants are some of the major sources of antioxidant compounds. Some studies have also shown that raw plant materials and single constituents like [6] gingerol have the ability to protect against lipid peroxidation. It is known that natural products and various medicinal plants change the blood metabolites of poultry when used as dietary supplements [95]. Still, the exact mechanisms through which blood metabolites are altered are not known. Research conducted with mice and ginger dietary supplements postulated that (E)-8 beta, 17-epoxyllabed-12-ene-15, 16-dial, a compound isolated from ginger, interferes with cholesterol biosynthesis in the liver homogenates of hypercholesterolaemic mice, causing their reduction. Srinivasan and Sambaiah [96] reported that feeding laboratory animals with natural plant products significantly elevated the activity of hepatic cholesterol 7-alpha-hydroxylase, which is a rate-limiting enzyme in the biosynthesis of bile acids and stimulates the conversion of cholesterol to bile acids, leading to the excretion of cholesterol from the body.

3. Bee Products and Their Effects on Health Status and Poultry Productivity

Honeybee products are natural substances synthesized in honeybees (*Apis mellifera*) [97–100]. These products have unique structures rich with active components of enzymes and peptides that have several pharmaceutical characteristics besides their high nutritional value and their significant impact on poultry's physiological and productive performance [101]. These kinds of nutritional and therapy alternatives have been used for centuries, mainly in China and Egypt [102]. According to the results of Babaei et al. [49], the critical active nutrient components of bee products may increase body weight gain, body weight, lymphoid organ weight, and antibody titer of Japanese quails. Rabie et al. [60] indicated that it is imperative to add bee products such as propolis (400 mg/kg diet) or bee pollen (2000 mg/kg) to broiler diets and bee venom (2 mg/L) to broiler's water as alternatives to antibiotics in poultry production. These additives can improve physiological performance, productivity, meat quality, and the development of the poultry immunity system due to their high levels of active enzymes, essential amino acids, vitamins, and minerals, and antimicrobial and immuno-stimulant activities of bee products [103]. Bee products can also increase the fertility of animals by enhancing the cryopreservation of gametes and fertilization [104]. The bioavailability of bee products is greater than that of artificially produced preparations. Research on the use of bee products as nutritional supplements for poultry has generally shown their positive impact on health and productivity [105,106]. A wide range of research has concentrated on the propolis supplementation of layers and broilers in various conditions and age groups. Research has concerned primarily the immune response of the birds, physiological parameters, and weight gain in broilers, as well as the parameters of egg-laying activity and quality of eggs [107].

3.1. Honey

Honey is a naturally sweet liquid produced by honeybees from nectar or plant secretions [108]. Therefore, several natural substances, such as antioxidants, are transferred from plants and accumulated in this product. The bioactive components of honey, including its chemical properties and physical characteristics, make it one of the most important natural products nowadays [109]. Chemically, it is composed of simple sugars: fructose, glucose; minerals: calcium, phosphorus; vitamins: ascorbic acid, riboflavin; enzymes; organic acid: gluconic, butyric; flavonoids and other phenolic and aromatic substances [110]. For its nutritional and therapeutic properties, honey is considered a vital drinking water additive for poultry [111]. The supplementation with honey into the drinking water (20 to 60 g/L) for broiler chickens during the summer seasons can improve some stress indices by more than 10%, body mass over 6%, and immunity over 2%. Haščík et al. [111] have conducted a small study on a total of 240 Ross and 308 broiler chickens to investigate the influence of bee pollen extract, propolis extract, and probiotics on the amino acid profile of chicken meat. Haščík et al. [111] stated that in addition, these bee products did not improve or show any adverse effects in the chicken meat amino acid profile. Another study was performed by Abioja et al. [112] on broiler chickens, with the aim to investigate the growth, mineral deposition, and physiological responses of chickens offered honey in drinking water during the hot-dry season. Investigations have shown that honey did not affect growth but might improve the chickens' welfare when offered up to 20 mL per litre water from d28 to d56 during hot periods. Adekunle et al. [113] investigated the addition of honey to the water of laying pullets through the hot–dry season. The trial was conducted on a total of 120 Isa Brown layers at 28 weeks old in a 16-week experiment supplemented with 10 and 20 mL of honey per 1 L of drinking water. The results of this trial have shown that heart rate was significantly ($p < 0.05$) lower in 20 mL supplemented hens (300.9 $\pm$ 1.70 bpm) compared to 0 mL supplemented hens (313 $\pm$ 1.70 bpm). Hens supplied 20 mL had a significantly ($p < 0.05$) higher lymphocyte count (50.6 $\pm$ 0.79 %) while hens that received a lower concentration of honey had a significantly ($p < 0.05$) lower basophil count (3.1 $\pm$ 0.39%). Adekunle et al. [113] concluded that the use of honey in drinking water reduced heart rate and basophil count.

3.2. Royal Jelly

Royal jelly is one of the popular bee products widely used as natural food for humans and animals due to its high content of essential nutrients [114]. It is an excellent source of B vitamins, vitamin C, folic acid, and phenolic acids [115]. Royal jelly is also a good source of minerals [116]. It has several important biological functions in the living being, including its effects as an antioxidant, immuno-stimulant, and growth promoter [117]. The antioxidant activity of royal jelly is primarily due to the presence of polyphenolic compounds [118]. It can be used in poultry production to improve the growth, gut health, and immune response and produce high-quality and safe poultry meat [119]. Previous studies focused on supplementing with royal jelly (10 to 200 mg/kg) in poultry diets have shown a significant increase in body weight (7%), egg production (10%), semen quality, and immune levels, as well as producing organic products [120]. Besides, El-Tarabany [52] conducted an experiment to clarify the influence of royal jelly on behavioural patterns, feather cover, egg quality, and some blood haematological indices in laying hens (58–64 weeks of age). Pure royal jelly injected in a concentration of 100 mg/kg, and 200 mg/kg, was used. Results have revealed that the feed consumption and water drinking in the 200 mg/kg group were significantly higher than the control and 100 mg/kg groups. Furthermore, the aggressive pecks, feather pecks, and threatening behaviour in the 200 mg/kg group were significantly lower than the control and 100 mg/kg groups. The effect of royal jelly, propolis, honey, and bee pollen was investigated in comparison to virginiamycin in regards to the performance and immune system of Japanese quail in research by Babaei et al. [49]. An in vivo experiment including a control, ethanolic extract of propolis 1000 and 5000 mg/kg, pollen powder 1000 and 5000 mg/kg, royal jelly 100 mg/kg, honey 22 g/L of drinking water, and virginiamycin 150 mg/kg were used in a trial for 42 days.

3.3. Bee Venom

Bee venom is also one of the bee products synthesized in the venom gland of honeybees and has several pharmaceutical and medical characteristics [121]. It consists of various substances, including peptides and enzymes, whereas melittin is the most effective component [122]. It also contains other essential substances, such as apamin and adolapin, which have various medicinal effects such as anti-inflammatory and antibacterial [123]. Therefore, bee venom could be added to the animal diet to provide productive and health benefits, including preventing and treating disease [124]. El-Hanoun et al. [125] and Elkomy et al. [124] reported that bee venom could also be an effective and safe alternative for poultry production instead of artificial sexual stimulants, which could harm consumer's health. It can improve reproductive efficiency, serum quality, and antioxidant status of broiler chickens and their immune response by using small doses (0.1 to 0.5 mg/kg) [121]. Rabie et al. [60] have conducted trials to evaluate the effects of propolis, bee pollen, and bee-venom as eco-friendly alternatives on the productive and physiological performance of broiler chickens. The chicks fed diets containing propolis (200 or 400 mg/kg diet), bee-venom (2 mg/L water), or bee-pollen (2 g/kg diet) showed significantly lower plasma cholesterol and LDL-cholesterol concentration compared to the control and Biox-Y®treatments. Consequently, propolis (200 or 400 mg/kg diet) and bee-pollen (2 g/kg diet) treatments had significantly higher plasma HDL cholesterol concentrations than the control treatment. Broilers fed propolis (200 or 400 mg/kg diet) for six weeks had significantly lower serum AST and ALT concentrations compared to the control treatment [60].

3.4. Bee Pollen

Bee pollen is a fresh form of feed additive and is composed of a range of nutritional and bioactive compounds [30]. Bee pollen has the ability to improve immunity, promote growth, protect intestinal tract health, and improve the quality and safety of poultry products. Also, bee pollen is a natural mixture of pollen grains collected by bee workers and then mixed with nectar and other gland secretions containing active components such as a β-glycosidase enzyme [126]. It has been used as natural growth and health promotion in poultry production due to its high content of amino acids, vitamins, and minerals [127]. Moreover, it is rich in polyphenols and tannins that act as protective agents and potent antioxidants, essential for improving animal health and immunity [128]. Abdelnour et al. [129] concluded that bee pollen could improve the animal's productive performance, including body weight gain. It can reduce daily feed intake by improving the feed conversion ratio [65] while increasing body weight by increasing the small intestinal villi surface of the duodenum, jejunum, and ileum.

Additionally, biochemical blood profiles, kidney functions, immunological parameters, antioxidant status, carcass traits, and meat quality have been improved remarkably by using bee pollen as an addition to broiler diets at the level of 400 to 800 mg/kg [129]. Bee pollen's underlying mechanism for improving carcass and meat quality could be attributed to decreased fat deposition and an increase in amino acid deposition [130]. The dietary addition of bee pollen in the concentration of 400 mg/kg have improved the amino acid profile of breast and thigh muscles, with significantly increased concentrations of tyrosine in the breast and significantly decreased concentrations of proline in breast and thigh muscles, respectively [131]. Trembecká et al. [132] have investigated the influence of bee pollen on the chemical and sensory characteristics of chicken breast meat. Results have shown that diet did not affect the chemical characteristics of chicken meat, except for supplementation with bee pollen and probiotics, which resulted in increased fat content. Further research has shown that the carcass yield values were significantly higher ($p = 0.038$), and the drip loss values were significantly lower ($p = 0.003$) in the experimental groups in comparison to the control group when bee pollen and propolis was used as dietary supplements. In addition, there was a statistically significant difference in b* skin colour ($p = 0.017$) and b* chicken breast muscle colour ($p < 0.001$) between the groups of chickens. The study showed that dietary supplementation with propolis and bee pollen has a significant positive effect on the quality of chicken meat [133]. Moreover, supplementation with bee pollen under stress conditions has been shown to reduce oxidative stress markers and improve the animal's antioxidant system [134]. Research has revealed that bee pollen as a natural antioxidant can be used as a supplementation in laying hen diets (1000 to 2000 mg/kg) to improve their production performance, egg production by 4.5%, and enhance egg yolk quality with a decreased total cholesterol concentration by 45%, thereby producing healthy products for consumers [135]. Furthermore, it is recommended to use bee pollen in roosters' diets (1000 mg/kg) to improve the ejaculate volume, progressive motility, live sperm, concentrate per ejaculate, and decrease sperm abnormalities [136].

3.5. Bee Propolis

Bee propolis is a natural resinous substance rich in active enzymes, and it has been recognized for its medicinal and therapeutic properties [137,138]. Aromatic compounds, flavonoids, and phenolics are the core components responding to the biological functions of propolis [139]. Working on propolis as one of the bee products used in poultry production, researchers have reported that it could be effectively added to the Japanese quail diet (500 to 4000 mg/kg) to improve growth performance and egg quality. It could optimize the lipid profile in egg yolk and reduce the total cholesterol concentration by more than 3.5%, respectively [140]. Moreover, it positively affects the antioxidative status of poultry, especially under summer conditions [141]. The authors found that propolis supplementation improved the immune status of poultry and positively affected the plasma levels of calcium (23%), phosphorus (24%), and albumin (17%), in addition to normalizing the plasma levels of alanine aminotransferase (ALT) and cholesterol [142]. This additive has successfully maintained the performance and egg production of Japanese quail under heat stress conditions at adequate levels [64]. Pieroni et al. [143] have investigated the effects of dietary inclusion of other varieties of propolis, such as the green propolis, on productivity, egg quality, nutrient utilization, and duodenal morphology of 120 Japanese laying quail. Authors have pointed out that the inclusion of green propolis at 1500 ppm in the diet of Japanese laying quail improves productivity, egg quality, nutrient utilization ability, and duodenal morphology [143].

4. Botanicals and Their Effects on Health Status and Poultry Productivity

Medicinal plants possess many natural materials widely used as dietary supplements in poultry production [38]. Due to their nutritional and medicinal properties, they have many positive effects on poultry performance [144]. Phytochemicals, phenols, flavonoids, tannins, and essential oils are found in various botanicals, and their products have several activities inside the poultry body [144]. They can act as digestive enhancers and health promoters for various poultry species such as broiler chickens, Japanese quails, and laying hens [145]. As a result, they play several significant roles in increasing poultry productivity and immunity [146]. Because of the bio-security threats for human and animal health, which come from the escalating resistance of pathogens to antibiotics, the global need to remove antimicrobial growth promoters from animal diets are rising as a very important issue.

4.1. Fenugreek Seeds (Trigonella foenum L.)

Fenugreek seeds are known as medicinal seeds that have several therapeutic properties such as antibacterial and anti-inflammatory [147]. They are also rich in protein, fat, carbohydrates, and minerals and contain biotin and trimethylamine, which tend to stimulate the animals' appetite [148]. Studies by Tewari et al. [148] revealed that fenugreek seed dietary supplementation has a positive effect on the activities of the hepatic antioxidant defence enzymes in animals. Galactomannan is the major polysaccharide found in fenugreek seeds and represents approximately 50% of the seed weight [149]. Galactomannan in fenugreek has been identified as an anti-diabetic compound because of its ability to reduce blood glucose levels [150]. Gaikwad et al. [67] concluded that using fenugreek seed powder in amounts 1% to 1.5% as a natural supplement in the broiler diet can improve the feed conversion ratio and increase the live body weights. Furthermore, Yassin et al. [66] concluded that the inclusion of up to 3% fenugreek in broiler diets could improve the average daily gain and carcass characteristics of chickens. Several investigations have suggested that fenugreek seeds may have hypocholesterolemic activity and thus, may be efficient in the treatment of egg yolk cholesterol. Omri et al. [151] have evaluated the effect of dietary incorporation of 3% of fenugreek seed combined with 3% of linseed, 1% of garlic paste, and 0.078% of copper sulphate on laying performance, egg quality, and lipid profile. An experiment on a total of forty-four 41 week-old Novogen White laying hens lasted 42 days. The dietary addition of medicinal plant mix in this small trial showed positive results. The egg weight of hens was not affected by dietary additions of medicinal plants, but the egg yolk cholesterol and blood cholesterol were both reduced. Abdouli et al. [152] evaluated the effects of ground fenugreek seeds given to laying hens at 2, 4 or 6 g/hen/day on laying performance, egg quality characteristics, serum, and egg yolk cholesterol concentrations. A total of forty 52-week-old Lohmann White laying hens were fed for seven weeks in the experiment. The results of Abdouli et al. [152] showed that ground fenugreek seeds reduced blood serum cholesterol but did not affect egg yolk cholesterol. Authors have pointed out that ground fenugreek seeds given to Lohmann White laying hens at up to 6 g/hen/day had no effect on feed intake, laying production performance, and egg quality but reduced the hen's serum cholesterol [152]. A research study was conducted by Abbas [153] to find the effect of fenugreek, parsley, and sweet basil seeds as natural feed additives on broiler performance. Experimentation was performed on a total of 120 day-old chicks reared for 42 days. Chicks fed basil diets had significantly ($p < 0.05$) heavier body weight than those fed fenugreek diets. Carcass characteristics had no significant differences. A significant reduction occurred in serum cholesterol as compared to control diets. Abbas [153] suggested that the supplementation of broiler chick diets with (3 g/kg) basil or parsley seeds improve product performance.

4.2. Black Cumin (Nigella sativa L.)

Black cumin seeds are also known for their medicinal and pharmaceutical properties [154]. They contain alkaloids, volatile oils, antioxidants, and several pharmacologically active substances such as thymol [155]. Martínez Aispuro et al. [156] reported that black cumin promotes poultry production and health and plays a significant role as a natural antioxidant, immuno-stimulant, and anti-cancer agent. Black cumin seed supplementation in daily poultry diets (10 to 30 g/kg) improved nutrient utilization, growth performance, and immune response and reduced the fatty acid content in poultry meat and eggs [157]. Rahman and Kim [158] indicated that black cumin supplementation could improve broiler chickens' production and meat quality by improving antioxidant activities and suppressing lipid peroxidation in meat. Boka et al. [159] performed a study to investigate the effects of different levels of black cumin seeds on performance, intestinal Escherichia coli count and morphology of jejunal epithelial cells in laying hens. A total of 100 Leghorn laying hens of 49 weeks old were used in an experiment where hens were supplemented with 0, 1, 2, and 3% of dietary black cumin. Based on the gain results, it has been concluded that supplementation with black cumin improves ($p < 0.05$) eggshell quality, Haugh unit, and feed conversion ratio. This improvement can be addressed as the increase ($p < 0.05$) in egg mass and contemporaneous decrease ($p < 0.01$) in feed consumption. Besides, regardless of supplementation level, dietary inclusion of black cumin decreased E. coli enumeration in ileal digesta and improved serum lipid profile and eggshell quality. A previous study conducted by Aydin et al. [160] with a total of 80 laying hens (Hyline-5 White) fed with a dietary addition of 1, 2, or 3% black cumin have shown a similar tendency.

4.3. Ginger (Zingiber officinale L.)

Ginger roots have been widely consumed as a herbal medicine with anti-cancer properties in several countries worldwide [161]. They contain essential volatile oils and other compounds, such as zingerone and gingerols, which can stimulate digestive enzymes and microbial activities and improve the antioxidative status of living beings [162]. Therefore, when Zhang et al. [70] have used a level of 5 g/kg ginger powder in broiler diets as nutritional supplementation, it led to the enhancement of antioxidant capacity and serum metabolites. Growing Japanese quails fed ginger supplemented diet (0.125 g/kg) obtained the best results in feed conversion ratios and humoral immunity. Moreover, ginger supplementation helped to optimize the lipid profile in blood serum and improve quails antioxidative status [163]. The use of ginger as an antibiotic replacement in livestock nutrition and welfare improvement may play a basic role in maximizing benefits and preserving poultry productiveness [164]. Oleforuh-Okoleh et al. [165] performed an experiment to evaluate the growth performance, haematological, and serum biochemical response of broiler chickens to aqueous extracts of ginger and garlic. The experiment lasted 56 days and was performed on a total of 80-day-old Marshal Strain broiler chickens supplemented with 50 mL of ginger and garlic in a 1:1 ratio mixture in drinking water. All the investigated parameters were significantly improved at the end of the experiment. On the other hand, a study using 144 one-day-old Arbor Acres broilers was conducted to assess the effects of dried ginger root processed to particle sizes of 300, 149, 74, 37, and 8.4 μm on growth performance, antioxidant status, and serum metabolites of broiler chickens by Zhang et al. [70]. The results of these investigations revealed that reducing the particle size of ginger powder linearly reduced ($p < 0.05$) cholesterol (d 21) and linearly increased ($p < 0.05$) glutathione peroxidase (d 21), total superoxide dismutase (d 42), and total protein (d 21 and 42). Supplementation of ginger at the level of 5 g/kg improved the antioxidant status of broilers, and the efficacy was enhanced as the particle size was reduced from 300 to 37 μm [70]. Other studies have been conducted to investigate the effect of ginger extract combined with citric acid on the tenderness of duck breast muscles. Even added as a marinade, ginger has expressed positive effects and significantly increased duck breast meat tenderness which could be attributed to various mechanisms such as increased myofibrillar fragmentation index and myofibrillar protein solubility [166].

4.4. Turmeric (Curcuma longa L.)

Turmeric is a rhizomatous herbaceous plant belonging to the ginger family [167]. It has been described as a natural polyphenol nutraceutical substance and is widely known for improving oxidative stress and fixing oxidative damage [168]. It can also be used in animal diets to mitigate heat stress [169]. Turmeric or curcumin can eliminate free radicals such as reactive oxygen species (ROS) and reactive nitrogen species (RNS) by activating glutathione (GSH), catalase, and superoxide dismutase (SOD), triggering better responses in the antioxidative mechanisms and inhibiting or neutralizing enzymes generators of ROS [170]. Therefore, it becomes an important antioxidant and medical additive in poultry diets, particularly in tropical regions, where high temperatures throughout the year can lead to reduced egg production, delayed growth, increased disease outbreaks, and mortality [171]. All these problems intensify poultry heat stress [172]. Sadeghi and Moghaddam [173] found that the addition of turmeric powder (0.5%) into broiler diet under heat stress increased serum concentrations of thyroxine (T_4) and triiodothyronine (T_3), the consumption of feed, the average daily gain, and decreased the feed conversion rate of birds. Furthermore, El-Maaty et al. [174] observed that turmeric powder (0.5 g/kg diet) could improve the final body weight (12%), the feed conversion ratio (19%), the digestibility of crude protein (5%), and the high-density lipoprotein (HDL) level (29%), as well as decrease the levels of creatinine, triglycerides, cholesterol (17%), low-density lipoprotein (LDL) (38%), during the heat stress conditions in broiler chickens. Algawany et al. [175] performed a study to evaluate the effects of turmeric for protection against alterations resulted from exposure to endosulfan in broiler chicks.

4.5. Thyme (Thymus vulgaris L.)

Thyme is one of the most commonly used herbs worldwide and is an aromatic medicinal plant, belongs to the Lamiaceae family [176]. It extensively uses in human food to add a distinctive aroma and flavour [177]. It is also used in animal diets as an alternative to synthetic drugs such as antibiotics to improve the product performance and the immunity of animals due to its antioxidant, antimicrobial, pharmaceutical, and therapeutic properties [178]. Dry thyme contains approximately 2% of the essential oil that is a mixture of monoterpenes, primarily thymol (2-isopropyl-5-methylphenol) and its phenol isomer carvacrol [179]. It also contains phenolics and some biphenolics and flavonoids, which have been shown to have an antioxidative effect and other benefits to poultry [180]. Thyme extracts are recommended to improve egg quality, particularly the fatty acid profile in yolk [181]. Yalçın et al. [182] found that a diet supplemented with dry thyme leaves (2%) has various beneficial effects on the physiological and productive performance, antioxidant status of laying hens, as well as having a positive effect on egg production quality such as decreasing the yolk cholesterol and total saturated fatty acids concentrations, and increasing the n-3 polyunsaturated fatty acids (PUFAs). It also increases the α-linolenic acid while decreasing palmitic acid percentage in the yolk [181]. Moreover, dietary thyme at 2% can decrease yolk malondialdehyde (MDA), blood serum cholesterol, and triglycerides levels, while the antibody titers against SRBC were increased [182]. Previous studies have shown that dietary thyme supplementation can increase the productivity of poultry (growth performance and egg production) and the humoral immune response without adverse effects on poultry [183–186]. Additionally, investigations have found that dietary thyme supplementation (0.1 to 1%) can play a role as a growth promoter in broiler diets to improve feed intake, conversion ratio, and body weight gain [187]. It can also improve the dressing percentage and meat quality of poultry. Its underlying mechanism may involve stimulating intramuscular fat and flavour amino deposition and changes in muscle fibre characteristics. Hashemipour et al. [188] reported that phytogenic products containing thymol and carvacrol improved performance, digestive enzyme activities, antioxidant enzyme activities, immune response, and retarded lipid oxidation in broiler chickens. Pournazari et al. [189] performed an investigation to evaluate the effects of the prebiotic, probiotic, and thyme essential oil on growth, organ and carcass traits, and hematology of Ross broiler chicks during 42 days on a total of 560 broilers. Results showed an increase in body weight gain and feed intake when thyme essential oil at 1 g/kg was used [189].

Additionally, thyme is rich in flavonoids, which can increase vitamin C activity, which serves as an antioxidant and, consequently, enhances immune functions [190]. Thus, it plays a vital role in reducing stress on poultry, particularly during the summer months, and increasing feed intake, improving feed metabolism, and decreasing weight loss in poultry, mainly due to heat stress [191]. Essential oils found in thyme can improve the productive performance of the animal by increasing the secretion of digestive enzymes and enhance the use of nutrients through enhanced liver function [192]. El-Ghousein and Al-Beitawi [193] found that the antibacterial activity of thyme could be associated with improved broiler productivity. In addition, Hernandez et al. [194] observed that broilers fed with a thyme-supplemented diet had improved their productive performance, which may be attributed to the improvement of apparent total faecal digestibility and crude protein digestibility. Another study was performed by Nouri [195] to evaluate the effects of chitosan nano-encapsulating mint, thyme, and cinnamon essential oils used in the diet for performance, immune responses and intestinal bacteria population in broiler chickens. A total of 600 mixed-sex, 1-d-old Ross 308 broiler chicks were used in the experiment, which lasted 42 days, during which chickens were supplemented with 0.025%, 0.04%, and 0.055% essential oils, respectively to starter, grower, and finisher diets. Results showed that thyme essential oil improved traits in broiler chickens [195].

5. Conclusions

Bee and botanical products are biologically active substances found in natural products. Whether they are from animal or plant sources, they are essential materials for maintaining better growth performance for animals such as poultry and enhancing their health and welfare.

The present review has focused on bee and botanical products' potential in poultry production as natural growth promoters, antioxidants, anti-inflammatory, and antimicrobial agents' immuno-stimulants, and sexual-stimulants. It has presented the current state of knowledge on the nutritional and therapeutic properties of the most important bee and natural botanical products in poultry diets and summarized the results of previous studies concerning dietary supplements in intensive and modern poultry production.

Bee and botanical supplements can increase the bodyweight of broilers and the egg production of laying hens by approximately 7 and 10% and enhance meat and egg quality by more than 25%.

Many results recommend that poultry have bee and botanical products as additions in their diets as natural alternatives to any artificial chemical material such as antibiotics. These natural substances can be added to the poultry diet individually or mixed to improve several physiological functions of poultry that have the most significant impact on the growth performance, reproductive efficiency, immune-response, and productivity of poultry, which, in turn, will lead to improved poultry welfare and produce healthy and organic products.

Keeping in mind that bee products are rich in several chemical bioactive compounds like polyphenols, steroids, terpenoids, amino acids, and inorganic compounds, and possesses many biological properties, including antiviral, antioxidant, antibacterial, hepatoprotective, antifungal and immuno-stimulating activities, further research on finding the ideal techniques for the practical application of bee products on a commercial level in poultry farming is still necessary and should be encouraged.

Also, botanical products represent a source of very valuable bioactive compounds with a proven impact on the health, performance, and welfare of poultry. These factors can be influenced by dosage, type, and levels of bioactive compounds, so further investigation should be focused on dosage and depend on trials and correlation effects with bee products to get deeper knowledge and mechanisms of action which will scientifically explain positive results on poultry performance and health support. It is well-known that botanicals have a positive influence as a growth enhancer, antioxidant, immunomodulator, antibacterial. Further, we suggest that the use of botanicals combined with bee products could be used for the alleviation of oxidative stress in farm animals and poultry, thus may be used successfully to overcome stress, but future research in this field is still needed.

Author Contributions: Conceptualization, N.P. and M.K.; methodology, S.V.; software, I.M.; investigation, N.P.; data curation, E.L.; writing—original draft preparation, E.L.; writing—review and editing, N.P.; visualization, N.P. All authors have read and agreed to the published version of the manuscript.

Funding: This research received no external funding.

Institutional Review Board Statement: Not applicable.

Informed Consent Statement: Not applicable.

Data Availability Statement: Data is contained within the article.

References

1. Alonso, M.E.; González-Montaña, J.R.; Lomillos, J.M. Consumers' Concerns and Perceptions of Farm Animal Welfare. *Animals* **2020**, *10*, 385. [CrossRef] [PubMed]
2. Su, Q.; Ganesh, S.; Moreno, M.; Bommireddy, Y.; Gonzalez, M.; Reklaitis, G.V.; Nagy, Z.K. A perspective on Quality-by-Control (QbC) in pharmaceutical continuous manufacturing. *Comput. Chem. Eng.* **2019**, *125*, 216–231. [CrossRef]
3. Tasić, S. Geographical and Economic Performance of Organic Agriculture and Its Impact on the Stability of Gastronomy Tourism in Serbia. *Oditor* **2018**, *4*, 38–51. [CrossRef]
4. Nowakiewicz, A.; Zięba, P.; Gnat, S.; Matuszewski, Ł. Last Call for Replacement of Antimicrobials in Animal Production: Modern Challenges, Opportunities, and Potential Solutions. *Antibiotics* **2020**, *9*, 883. [CrossRef] [PubMed]
5. Puvača, N.; Frutos, R.D.L. Antimicrobial Resistance in *Escherichia coli* Strains Isolated from Humans and Pet Animals. *Antibiotics* **2021**, *10*, 69. [CrossRef] [PubMed]
6. Yadav, S.; Jha, R. Strategies to modulate the intestinal microbiota and their effects on nutrient utilization, performance, and health of poultry. *J. Anim. Sci. Biotechnol.* **2019**, *10*, 1–11. [CrossRef]
7. Puvača, N.; Čabarkapa, I.; Bursić, V.; Petrović, A.; Aćimović, M. Antimicrobial, Antioxidant and Acaricidal Properties of Tea Tree (*Melaleuca alternifolia*). *J. Agron. Technol. Eng. Manag.* **2018**, *1*, 29–38.
8. Marotta, F.; Garofolo, G.; Di Marcantonio, L.; Di Serafino, G.; Neri, D.; Romantini, R.; Sacchini, L.; Alessiani, A.; Di Donato, G.; Nuvoloni, R.; et al. Antimicrobial resistance genotypes and phenotypes of Campylobacter jejuni isolated in Italy from humans, birds from wild and urban habitats, and poultry. *PLoS ONE* **2019**, *14*, e0223804. [CrossRef]
9. Lekshmi, M.; Ammini, P.; Kumar, S.; Varela, M.F. The Food Production Environment and the Development of Antimicrobial Resistance in Human Pathogens of Animal Origin. *Microorganisms* **2017**, *5*, 11. [CrossRef]
10. Paphitou, N.I. Antimicrobial resistance: Action to combat the rising microbial challenges. *Int. J. Antimicrob. Agents* **2013**, *42*, S25–S28. [CrossRef]
11. Ferri, M.; Ranucci, E.; Romagnoli, P.; Giaccone, V. Antimicrobial resistance: A global emerging threat to public health systems. *Crit. Rev. Food Sci. Nutr.* **2015**, *57*, 2857–2876. [CrossRef] [PubMed]
12. Ricke, S.C. Prebiotics and Alternative Poultry Production. *Poult. Sci.* **2021**, *100*, 101174. [CrossRef] [PubMed]
13. Ghosh, S.; Playford, R.J. Bioactive natural compounds for the treatment of gastrointestinal disorders. *Clin. Sci.* **2003**, *104*, 547–556. [CrossRef] [PubMed]
14. Isman, M.B. Botanical Insecticides, Deterrents, and Repellents in Modern Agriculture and an Increasingly Regulated World. *Annu. Rev. Èntomol.* **2006**, *51*, 45–66. [CrossRef] [PubMed]

15. Eteraf-Oskouei, T.; Najafi, M. Traditional and Modern Uses of Natural Honey in Human Diseases: A Review. *Iran. J. Basic Med. Sci.* **2013**, *16*, 731–742. [PubMed]
16. Windisch, W.; Schedle, K.; Plitzner, C.; Kroismayr, A. Use of phytogenic products as feed additives for swine and poultry. *J. Anim. Sci.* **2008**, *86*, E140–E148. [CrossRef]
17. Gheisar, M.M.; Kim, I.H. Phytobiotics in poultry and swine nutrition—A review. *Ital. J. Anim. Sci.* **2017**, *17*, 92–99. [CrossRef]
18. Lee, M.T.; Lin, W.C.; Yu, B.; Lee, T.-T. Antioxidant capacity of phytochemicals and their potential effects on oxidative status in animals—A review. *Asian-Australas. J. Anim. Sci.* **2016**, *30*, 299–308. [CrossRef] [PubMed]
19. Pandey, A.K.; Kumar, P.; Saxena, M.J. Feed Additives in Animal Health. In *Nutraceuticals in Veterinary Medicine*; Gupta, R.C., Srivastava, A., Lall, R., Eds.; Springer International Publishing: Cham, Switzerland, 2019; pp. 345–362, ISBN 978-3-030-04624-8.
20. Cabarkapa, I.; Puvača, N.; Popović, S.; Čolović, D.; Kostadinović, L.; Tatham, E.K.; Lević, J. Aromatic plants and their extracts pharmacokinetics and in vitro/in vivo mechanisms of action. In *Feed Additives*; Elsevier: Amsterdam, The Netherlands, 2020; pp. 75–88, ISBN 978-0-12-814700-9.
21. Popović, S.; Kostadinović, L.; Đuragić, O.; Aćimović, M.; Čabarkapa, I.; Puvača, N.; Pelić, D.L. Influence of Medicinal Plants Mixtures (*Artemisia absinthium, Thymus vulgaris, Menthae piperitae and Thymus serpyllum*) In Broilers Nutrition on Biochemical Blood Status. *J. Agron. Technol. Eng. Manag.* **2018**, *1*, 91–98.
22. Singh, A.K.; Cabral, C.; Kumar, R.; Ganguly, R.; Rana, H.K.; Gupta, A.; Lauro, M.R.; Carbone, C.; Reis, F.; Pandey, A.K. Beneficial Effects of Dietary Polyphenols on Gut Microbiota and Strategies to Improve Delivery Efficiency. *Nutrients* **2019**, *11*, 2216. [CrossRef]
23. Cross, D.; McDevitt, R.M.; Hillman, K.; Acamovic, T. The effect of herbs and their associated essential oils on performance, dietary digestibility and gut microflora in chickens from 7 to 28 days of age. *Br. Poult. Sci.* **2007**, *48*, 496–506. [CrossRef]
24. Yang, Y.; Iji, P.; Choct, M. Dietary modulation of gut microflora in broiler chickens: A review of the role of six kinds of alternatives to in-feed antibiotics. *World's Poult. Sci. J.* **2009**, *65*, 97–114. [CrossRef]
25. Mahady, G.B. Medicinal Plants for the Prevention and Treatment of Bacterial Infections. *Curr. Pharm. Des.* **2005**, *11*, 2405–2427. [CrossRef]
26. Pliego, A.B.; Tavakoli, M.; Khusro, A.; Seidavi, A.; Elghandour, M.M.M.Y.; Salem, A.Z.M.; Márquez-Molina, O.; Rivas-Caceres, R.R. Beneficial and adverse effects of medicinal plants as feed supplements in poultry nutrition: A review. *Anim. Biotechnol.* **2020**, 1–23. [CrossRef]
27. Raza, A.; Muhammad, F.; Bashir, S.; Anwar, M.; Awais, M.; Akhtar, M.; Aslam, B.; Khaliq, T.; Naseer, M. Antiviral and immune boosting activities of different medicinal plants against Newcastle disease virus in poultry. *World's Poult. Sci. J.* **2015**, *71*, 523–532. [CrossRef]
28. Seidavi, A.R.; Laudadio, V.; Khazaei, R.; Puvača, N.; Selvaggi, M.; Tufarelli, V. Feeding of black cumin (*Nigella sativa* L.) and its effects on poultry production and health. *World's Poult. Sci. J.* **2020**, *76*, 346–357. [CrossRef]
29. McMurray, R.; Ball, M.; Tunney, M.; Corcionivoschi, N.; Situ, C. Antibacterial Activity of Four Plant Extracts Extracted from Traditional Chinese Medicinal Plants against *Listeria monocytogenes, Escherichia coli*, and *Salmonella enterica* subsp. *enterica* serovar Enteritidis. *Microorganisms* **2020**, *8*, 962. [CrossRef] [PubMed]
30. Haščík, P.; Pavelkova, A.; Bobko, M.; Trembecká, L.; Elimam, I.; Capcarova, M. The effect of bee pollen in chicken diet. *World's Poult. Sci. J.* **2017**, *73*, 643–650. [CrossRef]
31. Oke, O.E.; Sorungbe, F.O.; Abioja, M.O.; Oyetunji, O.; Onabajo, A.O. Effect of different levels of honey on physiological, growth and carcass traits of broiler chickens during dry season. *Acta Agric. Slov.* **2016**, *108*, 45–53. [CrossRef]
32. Abioja, M.O.; Adekunle, M.O.; Abiona, J.A.; Sodipe, O.G.; Jegede, A.V. Laying Performance, Survival Rate, Egg Quality and Shell Characteristics in Laying Pullets Offered Honey in Drinking Water during Hot Season. *Agric. Trop. Subtrop.* **2016**, *49*, 12–19. [CrossRef]
33. Malayoğlu, H.B.; Baysal, Ş.E.N.A.Y.; Misirlioğlu, Z.; Polat, M.; Yilmaz, H.; Turan, N. Effects of oregano essential oil with or without feed enzymes on growth performance, digestive enzyme, nutrient digestibility, lipid metabolism and immune response of broilers fed on wheat–soybean meal diets. *Br. Poult. Sci.* **2010**, *51*, 67–80. [CrossRef]
34. Das, L.; Bhaumik, E.; Raychaudhuri, U.; Chakraborty, R. Role of nutraceuticals in human health. *J. Food Sci. Technol.* **2011**, *49*, 173–183. [CrossRef]
35. Kennedy, D.O.; Wightman, E.L. Herbal Extracts and Phytochemicals: Plant Secondary Metabolites and the Enhancement of Human Brain function. *Adv. Nutr.* **2011**, *2*, 32–50. [CrossRef] [PubMed]
36. Santhosha, S.; Jamuna, P.; Prabhavathi, S. Bioactive components of garlic and their physiological role in health maintenance: A review. *Food Biosci.* **2013**, *3*, 59–74. [CrossRef]
37. Isman, M.B.; Grieneisen, M.L. Botanical insecticide research: Many publications, limited useful data. *Trends Plant Sci.* **2014**, *19*, 140–145. [CrossRef] [PubMed]
38. Puvača, N.; Lika, E.; Tufarelli, V.; Bursić, V.; Pelić, D.L.; Nikolova, N.; Petrović, A.; Prodanović, R.; Vuković, G.; Lević, J.; et al. Influence of Different Tetracycline Antimicrobial Therapy of Mycoplasma (*Mycoplasma synoviae*) in Laying Hens Compared to Tea Tree Essential Oil on Table Egg Quality and Antibiotic Residues. *Foods* **2020**, *9*, 612. [CrossRef]
39. Arenas-Jal, M.; Suñé-Negre, J.M.; Pérez-Lozano, P.; García-Montoya, E. Trends in the food and sports nutrition industry: A review. *Crit. Rev. Food Sci. Nutr.* **2019**, *60*, 2405–2421. [CrossRef] [PubMed]

40. Sugiharto, S. Role of nutraceuticals in gut health and growth performance of poultry. *J. Saudi Soc. Agric. Sci.* **2016**, *15*, 99–111. [CrossRef]

41. Duda-Chodak, A.; Tarko, T.; Satora, P.; Sroka, P. Interaction of dietary compounds, especially polyphenols, with the intestinal microbiota: A review. *Eur. J. Nutr.* **2015**, *54*, 325–341. [CrossRef]

42. Farag, M.A.; Abdelwareth, A.; Sallam, I.; Shorbagi, M.; Jehmlich, N.; Fritz-Wallace, K.; Schäpe, S.S.; Rolle-Kampczyk, U.; Ehrlich, A.; Wessjohann, L.A.; et al. Metabolomics reveals impact of seven functional foods on metabolic pathways in a gut microbiota model. *J. Adv. Res.* **2020**, *23*, 47–59. [CrossRef]

43. Gowd, V.; Karim, N.; Shishir, M.R.I.; Xie, L.; Chen, W. Dietary polyphenols to combat the metabolic diseases via altering gut microbiota. *Trends Food Sci. Technol.* **2019**, *93*, 81–93. [CrossRef]

44. Possemiers, S.; Bolca, S.; Verstraete, W.; Heyerick, A. The intestinal microbiome: A separate organ inside the body with the metabolic potential to influence the bioactivity of botanicals. *Fitoterapia* **2011**, *82*, 53–66. [CrossRef] [PubMed]

45. Aluko, R. Bioactive Carbohydrates. In *Functional Foods and Nutraceuticals*; Aluko, R.E., Ed.; Food Science Text Series; Springer: New York, NY, USA, 2012; pp. 3–22. ISBN 978-1-4614-3480-1.

46. Khan, N.; Abbasi, A.M.; Dastagir, G.; Nazir, A.; Shah, G.; Shah, M.M.; Shah, M.H. Ethnobotanical and antimicrobial study of some selected medicinal plants used in Khyber Paktunkhwa (KPK) as a potential source to cure infectious diseases. *BMC Complement. Altern. Med.* **2014**, *14*, 122. [CrossRef]

47. Ghanima, M.M.A.; Elsadek, M.F.; Taha, A.E.; El-Hack, M.E.A.; Alagawany, M.; Ahmed, B.M.; Elshafie, M.M.; El-Sabrout, K. Effect of Housing System and Rosemary and Cinnamon Essential Oils on Layers Performance, Egg Quality, Haematological Traits, Blood Chemistry, Immunity, and Antioxidant. *Animals* **2020**, *10*, 245. [CrossRef]

48. Puvača, N. Bioactive Compounds in Selected Hot Spices and Medicinal Plants. *J. Agron. Technol. Eng. Manag.* **2018**, *1*, 8–17.

49. Babaei, S.; Rahimi, S.; Torshizi, M.A.K.; Tahmasebi, G.; Miran, S.N.K. Effects of propolis, royal jelly, honey and bee pollen on growth performance and immune system of Japanese quails. *Vet. Res. Forum* **2016**, *7*, 13–20. [PubMed]

50. Babaei, S.; Rahimi, S.; Karimi Torshizi, M.A.; Tahmasebi, G.H.; Khaleghi Miran, S.N. Effect of Honey, Royal Jelly and Bee Pollen on Performance, Immune System and Blood Parameters in Japanese Quail. *Anim. Prod.* **2015**, *17*, 311–320. [CrossRef]

51. Pavelková, A.; Haščí k, P.; Capcarová, M.; Kalafová, A.; Hanusová, E.; Tkáčová, J.; Bobko, M.; Čuboň, J.; Čech, M.; Kačániová, M. Meat performance of Japanese quails after the application of bee bread powder. *Potravin. Slovak J. Food Sci.* **2020**, *14*, 735–743. [CrossRef]

52. El-Tarabany, M.S. Effect of Royal Jelly on behavioural patterns, feather quality, egg quality and some haematological parameters in laying hens at the late stage of production. *J. Anim. Physiol. Anim. Nutr.* **2017**, *102*, e599–e606. [CrossRef] [PubMed]

53. Zhu, A.; Zhang, K.; Wang, J.; Bai, S.; Zeng, Q.; Peng, H.; Ding, X. Effect of different concentrations of neohesperidin dihydrochalcone on performance, egg quality, serum biochemistry and intestinal morphology in laying hens. *Poult. Sci.* **2021**, *100*, 101097. [CrossRef]

54. Kim, D.-H.; Han, S.-M.; Choi, Y.-S.; Kang, H.-K.; Lee, H.-G.; Lee, K.-W. Effects of Dietary Bee Venom on Serum Characteristic, Antioxidant Activity and Liver Fatty Acid Composition in Broiler Chickens. *Korean J. Poult. Sci.* **2019**, *46*, 39–46. [CrossRef]

55. Han, S.M.; Lee, K.G.; Yeo, J.H.; Oh, B.Y.; Kim, B.S.; Lee, W.; Baek, H.J.; Kim, S.T.; Hwang, S.J.; Pak, S.C. Effects of honeybee venom supplementation in drinking water on growth performance of broiler chickens. *Poult. Sci.* **2010**, *89*, 2396–2400. [CrossRef] [PubMed]

56. Abuoghaba, A.A.-K. Egg Production, Egg Quality Traits and Some Hematological Parameters of Sinai Chicken Strain Treated With Different Levels of Bee Pollen. *Egypt. Poult. Sci. J.* **2018**, *38*, 427–438.

57. Desoky, A.; Kamel, N. Egg Production Performance, Blood Biochemical and Immunological Response of Laying Japanese Quail Fed Diet Supplemented with Propolis and Bee Pollen. *Egypt. J. Nutr. Feeds* **2018**, *21*, 549–557. [CrossRef]

58. De Oliveira, M.; Da Silva, D.; Loch, F.; Martins, P.; Dias, D.; Simon, G. Effect of bee pollen on the immunity and tibia characteristics in broilers. *Braz. J. Poult. Sci.* **2013**, *15*, 323–327. [CrossRef]

59. Farag, S.A.; Rayes, T.E. Effect of Bee-pollen Supplementation on Performance, Carcass Traits and Blood Parameters of Broiler Chickens. *Asian J. Anim. Vet. Adv.* **2016**, *11*, 168–177. [CrossRef]

60. Rabie, A.H.; El-Kaiaty, A.M.; Hassan, M.S.; Stino, F.R. Influence of Some Honey Bee Products And A Growth Promoter Supplementation On Productive And Physiological Performance of Broiler Chickens. *Egypt. Poult. Sci. J.* **2018**, *38*, 513–531.

61. Abdel-Kareem, A.A.A.; El-Sheikh, T.M. Impact of supplementing diets with propolis on productive performance, egg quality traits and some haematological variables of laying hens. *J. Anim. Physiol. Anim. Nutr.* **2015**, *101*, 441–448. [CrossRef]

62. Çetin, E.; Silici, S.; Çetin, N.; Guclu, B.K. Effects of diets containing different concentrations of propolis on hematological and immunological variables in laying hens. *Poult. Sci.* **2010**, *89*, 1703–1708. [CrossRef]

63. Seven, P.T. The Effects of Dietary Turkish Propolis and Vitamin C on Performance, Digestibility, Egg Production and Egg Quality in Laying Hens under Different Environmental Temperatures. *Asian-Australas. J. Anim. Sci.* **2008**, *21*, 1164–1170. [CrossRef]

64. Mehaisen, G.M.K.; Desoky, A.A.; Sakr, O.G.; Sallam, W.; Abass, A.O. Propolis alleviates the negative effects of heat stress on egg production, egg quality, physiological and immunological aspects of laying Japanese quail. *PLoS ONE* **2019**, *14*, e0214839. [CrossRef]

65. de Oliveira, M.C.; de Souza, R.G.; Dias, D.M.B.; Gonçalves, B.N. Bee pollen improves productivity of laying Japanese quails. *Rev. Bras. Saúde Produção Anim.* **2020**, *21*. [CrossRef]

66. Yassin, M.; Nurfeta, A.; Banerjee, S. The Effect of Supplementing Fenugreek (*Trigonella foenum-graecum* L.) Seed Powder on Growth Performance, Carcass Characteristics and Meat Quality of Cobb 500 Broilers Reared on Conventional Ration. *Ethiop. J. Agric. Sci.* **2020**, *30*, 129–142.

67. Gaikwad, B.; Patil, R.; Padghan, P.; Shinde, S. Effect of Fenugreek (*Trigonella foenum-gracum* L.) Seed Powder as Natural Feed Additive on Performance and Blood Parameters of Broiler Chicks. *Int. J. Curr. Microbiol. Appl. Sci.* **2019**, *8*, 1147–1155. [CrossRef]

68. Laudadio, V.; Nasiri-Dehbaneh, M.; Bilal, R.M.; Qotbi, A.; Javandel, F.; Ebrahimi, A.; Seidavi, A.; Slozhenkina, M.; Gorlov, I.; Dunne, P.G.; et al. Effects of different levels of dietary black cumin (*Nigella sativa* L.) and fenugreek (*Trigonella foenum-graecum* L.) and their combination on productive traits, selected blood constituents, microbiota and immunity of broilers. *Anim. Biotechnol.* **2020**, 1–14. [CrossRef]

69. Hafeez, A.; Iqbal, S.; Sikandar, A.; Din, S.; Khan, I.; Ashraf, S.; Khan, R.; Tufarelli, V.; Laudadio, V. Feeding of Phytobiotics and Exogenous Protease in Broilers: Comparative Effect on Nutrient Digestibility, Bone Strength and Gut Morphology. *Agriculture* **2021**, *11*, 228. [CrossRef]

70. Zhang, G.; Yang, Z.; Wang, Y.; Yang, W.; Jiang, S.; Gai, G. Effects of ginger root (*Zingiber officinale*) processed to different particle sizes on growth performance, antioxidant status, and serum metabolites of broiler chickens. *Poult. Sci.* **2009**, *88*, 2159–2166. [CrossRef] [PubMed]

71. Ali, M.; Chand, N.; Khan, R.U.; Naz, S.; Gul, S. Anticoccidial effect of garlic (*Allium sativum*) and ginger (*Zingiber officinale*) against experimentally induced coccidiosis in broiler chickens. *J. Appl. Anim. Res.* **2019**, *47*, 79–84. [CrossRef]

72. Shewita, R.; Taha, A. Influence of dietary supplementation of ginger powder at different levels on growth performance, haematological profiles, slaughter traits and gut morphometry of broiler chickens. *S. Afr. J. Anim. Sci.* **2019**, *48*. [CrossRef]

73. Akbarian, A.; Golian, A.; Ahmadi, A.S.; Moravej, H. Effects of ginger root (*Zingiber officinale*) on egg yolk cholesterol, antioxidant status and performance of laying hens. *J. Appl. Anim. Res.* **2011**, *39*, 19–21. [CrossRef]

74. Zhao, X.; Yang, Z.; Yang, W.; Wang, Y.; Jiang, S.; Zhang, G. Effects of ginger root (*Zingiber officinale*) on laying performance and antioxidant status of laying hens and on dietary oxidation stability. *Poult. Sci.* **2011**, *90*, 1720–1727. [CrossRef]

75. Malekizadeh, M.; Moeini, M.M.; Ghazi, S. The Effects Of Different Levels Of Ginger (*Zingiber officinale* Rosc) And Turmeric (*Curcuma longa* Linn) Rhizomes Powder On Some Blood Metabolites And Production Performance Characteristics Of Laying Hens. *J. Agric. Sci. Technol.* **2012**, *14*, 127–134.

76. Gumus, H.; Oguz, M.N.; Bugdayci, K.E.; Oguz, F.K. Effects of sumac and turmeric as feed additives on performance, egg quality traits, and blood parameters of laying hens. *Rev. Bras. Zootec.* **2018**, *47*. [CrossRef]

77. Riasi, A. Production performance, egg quality and some serum metabolites of older commercial laying hens fed different levels of turmeric rhizome (*Curcuma longa*) powder. *J. Med. Plants Res.* **2012**, *6*. [CrossRef]

78. Sugiharto, S. Alleviation of heat stress in broiler chicken using turmeric (*Curcuma longa*)—A short review. *J. Anim. Behav. Biometeorol.* **2020**, *8*, 215–222. [CrossRef]

79. Akbarian, A.; Golian, A.; Kermanshahi, H.; Gilani, A.; Moradi, S. Influence of Turmeric Rhizome and Black Pepper on Blood Constituents and Performance of Broiler Chickens. *Afr. J. Biotechnol.* **2012**, *11*, 8606–8611. [CrossRef]

80. Khan, R.; Naz, S.; Javdani, M.; Nikousefat, Z.; Selvaggi, M.; Tufarelli, V.; Laudadio, V. The use of Turmeric (*Curcuma longa*) in poultry feed. *World's Poult. Sci. J.* **2012**, *68*, 97–103. [CrossRef]

81. Ahmadian, A.; Seidavi, A.; Phillips, C.J.C. Growth, Carcass Composition, Haematology and Immunity of Broilers Supplemented with Sumac Berries (*Rhus coriaria* L.) and Thyme (*Thymus vulgaris*). *Animals* **2020**, *10*, 513. [CrossRef] [PubMed]

82. Bolukbasi, S.C.; Erhan, M.K.; Ozkan, A. Effect of Dietary Thyme Oil and Vitamin E on Growth, Lipid Oxidation, Meat Fatty Acid Composition and Serum Lipoproteins of Broilers. *S. Afr. J. Anim. Sci.* **2006**, *36*, 189–196. [CrossRef]

83. Abdel-Wareth, A.; Kehraus, S.; Hippenstiel, F.; Südekum, K.-H. Effects of thyme and oregano on growth performance of broilers from 4 to 42 days of age and on microbial counts in crop, small intestine and caecum of 42-day-old broilers. *Anim. Feed Sci. Technol.* **2012**, *178*, 198–202. [CrossRef]

84. Beyzi, S.B.; Konca, Y.; Kaliber, M.; Sarıözkan, S.; Güçlü, B.K.; Aktuğ, E.; Şentürk, M. Effects of thyme essential oil and A, C, and E vitamin combinations to diets on performance, egg quality, MDA, and 8-OHdG of laying hens under heat stress. *J. Appl. Anim. Res.* **2020**, *48*, 126–132. [CrossRef]

85. Akbari, M.; Torki, M.; Kaviani, K. Single and combined effects of peppermint and thyme essential oils on productive performance, egg quality traits, and blood parameters of laying hens reared under cold stress condition (6.8 ± 3 °C). *Int. J. Biometeorol.* **2015**, *60*, 447–454. [CrossRef] [PubMed]

86. Arpášová, H.; Gálik, B.; Hrnčár, C.; Fik, M.; Herkeľ, R.; Pistová, V. The Effect of Essential Oils on Performance of Laying Hens. *Sci. Pap. Anim. Sci. Biotechnol.* **2015**, *48*, 8–14.

87. Abasi, O.; Daneshyar, M. Effect of Different Levels of Mentha Longifolia and Thymus Vulgaris Powders on Growth, Carcass Characteristics and Immune System of Japanese Quails. *Iran. Vet. J.* **2020**, *16*, 71–81. [CrossRef]

88. Khaksar, V.; van Krimpen, M.; Hashemipour, H.; Pilevar, M. Effects of Thyme Essential Oil on Performance, Some Blood Parameters and Ileal Microflora of Japanese Quail. *J. Poult. Sci.* **2012**, *49*, 106–110. [CrossRef]

89. Horáková, K.; Šovčíková, A.; Seemannová, Z.; Syrová, D.; Bušányová, K.; Drobná, Z.; Ferenčík, M. Detection of drug-induced, superoxide-mediated cell damage and its prevention by antioxidants. *Free Radic. Biol. Med.* **2001**, *30*, 650–664. [CrossRef]

90. Huang, S.-S.; Liu, S.-M.; Lin, S.-M.; Liao, P.-H.; Lin, R.-H.; Chen, Y.-C.; Chih, C.-L.; Tsai, S.-K. Antiarrhythmic effect of caffeic acid phenethyl ester (CAPE) on myocardial ischemia/reperfusion injury in rats. *Clin. Biochem.* **2005**, *38*, 943–947. [CrossRef] [PubMed]

91. Farooqui, T.; Farooqui, A.A. Aging: An important factor for the pathogenesis of neurodegenerative diseases. *Mech. Ageing Dev.* **2009**, *130*, 203–215. [CrossRef]
92. Song, Y.S.; Park, E.-H.; Hur, G.M.; Ryu, Y.S.; Lee, Y.S.; Lee, J.Y.; Kim, Y.M.; Jin, C. Caffeic acid phenethyl ester inhibits nitric oxide synthase gene expression and enzyme activity. *Cancer Lett.* **2002**, *175*, 53–61. [CrossRef]
93. Khayyal, M.T.; El-Ghazaly, M.A.; El-Khatib, A.; Hatem, A.M.; De Vries, P.J.F.; El-Shafei, S.; Khattab, M.M. A clinical pharmacological study of the potential beneficial effects of a propolis food product as an adjuvant in asthmatic patients. *Fundam. Clin. Pharmacol.* **2003**, *17*, 93–102. [CrossRef]
94. Khan, R.; Naz, S.; Nikousefat, Z.; Tufarelli, V.; Javdani, M.; Qureshi, M.S.; Laudadio, V. Potential applications of ginger (*Zingiber officinale*) in poultry diets. *World's Poult. Sci. J.* **2012**, *68*, 245–252. [CrossRef]
95. Saeid, J.M.; Mohamed, A.B.; Al-Baddy, M.A. Effect of Aqueous Extract of Ginger (*Zingiber officinale*) on Blood Biochemistry Parameters of Broiler. *Int. J. Poult. Sci.* **2010**, *9*, 944–947. [CrossRef]
96. Srinivasan, K.; Sambaiah, K. The effect of spices on cholesterol 7 alpha-hydroxylase activity and on serum and hepatic cholesterol levels in the rat. *Int. J. Vitam. Nutr. Res.* **1991**, *61*, 364–369.
97. Kolayli, S.; Keskin, M. Chapter 7—Natural bee products and their apitherapeutic applications. In *Studies in Natural Products Chemistry*; Rahman, A.U., Ed.; Bioactive Natural Products; Elsevier: Amsterdam, The Netherlands, 2020; Volume 66, pp. 175–196.
98. Lima, W.G.; Brito, J.C.M.; Nizer, W.S.D.C. Bee products as a source of promising therapeutic and chemoprophylaxis strategies against COVID-19 (SARS-CoV-2). *Phytother. Res.* **2020**, *35*, 743–750. [CrossRef] [PubMed]
99. El-Seedi, H.R.; Khalifa, S.A.; El-Wahed, A.A.; Gao, R.; Guo, Z.; Tahir, H.E.; Zhao, C.; Du, M.; Farag, M.A.; Musharraf, S.G.; et al. Honeybee products: An updated review of neurological actions. *Trends Food Sci. Technol.* **2020**, *101*, 17–27. [CrossRef]
100. Prodanović, R.; Ignjatijević, S.; Bošković, J. Innovative Potential of Beekeeping Production in AP Vojvodina. *J. Agron. Technol. Eng. Manag.* **2019**, *2*, 268–277.
101. Callaway, T.; Lillehoj, H.; Chuanchuen, R.; Gay, C. Alternatives to Antibiotics: A Symposium on the Challenges and Solutions for Animal Health and Production. *Antibiotics* **2021**, *10*, 471. [CrossRef]
102. Hashem, N.; Hassanein, E.; Simal-Gandara, J. Improving Reproductive Performance and Health of Mammals Using Honeybee Products. *Antioxidants* **2021**, *10*, 336. [CrossRef]
103. Haščík, P.; Pavelková, A.; Arpášová, H.; Čuboň, J.; Kačániová, M.; Kunová, S. The Effect of Bee Products And Probiotic On Meat Performance of Broiler Chickens. *J. Microbiol. Biotechnol. Food Sci.* **2021**, *9*, 88–92. [CrossRef]
104. Abdelnour, S.A.; El-Hack, M.E.A.; Alagawany, M.; Taha, A.E.; Elnesr, S.S.; Elmonem, O.M.A.; Swelum, A.A. Useful impacts of royal jelly on reproductive sides, fertility rate and sperm traits of animals. *J. Anim. Physiol. Anim. Nutr.* **2020**, *104*, 1798–1808. [CrossRef] [PubMed]
105. Galal, A.; Ahmed, A.; Ali, W.; El-Sanhoury, M.; Ahmed, H.E. Residual Feed Intake and its Effect on Cell-Mediated Immunity in Laying Hens Given Different Propolis Levels. *Int. J. Poult. Sci.* **2008**, *7*, 1105–1111. [CrossRef]
106. The Effect of Diet Propolis Supplementation on Ross Broiler Chicks Performance. *Int. J. Poult. Sci.* **2005**, *5*, 84–88. [CrossRef]
107. Humoral Immunity of Broilers is Affected by Oil Extracted Propolis (OEP) in the Diet. *Int. J. Poult. Sci.* **2005**, *4*, 414–417. [CrossRef]
108. Jiang, W.; Ying, M.; Zhang, J.; Cui, Z.; Chen, Q.; Chen, Y.; Wang, J.; Fang, F.; Shen, L. Quantification of major royal jelly proteins using ultra performance liquid chromatography tandem triple quadrupole mass spectrometry and application in honey authenticity. *J. Food Compos. Anal.* **2021**, *97*, 103801. [CrossRef]
109. Kanbur, E.D.; Yuksek, T.; Atamov, V.; Ozcelik, A.E. A comparison of the physicochemical properties of chestnut and highland honey: The case of Senoz Valley in the Rize province of Turkey. *Food Chem.* **2021**, *345*, 128864. [CrossRef] [PubMed]
110. Guo, N.; Zhao, L.; Zhao, Y.; Li, Q.; Xue, X.; Wu, L.; Escalada, M.G.; Wang, K.; Peng, W. Comparison of the Chemical Composition and Biological Activity of Mature and Immature Honey: An HPLC/QTOF/MS-Based Metabolomic Approach. *J. Agric. Food Chem.* **2020**, *68*, 4062–4071. [CrossRef]
111. Haščík, P.; Pavelková, A.; Tkáčová, J.; Čuboň, J.; Kačániová, M.; Habánová, M.; Mlyneková, E. The amino acid profile of broiler chicken meat after dietary administration of bee products and probiotics. *Biologia* **2020**, *75*, 1899–1908. [CrossRef]
112. Abioja, M.O.; Ogundimu, K.B.; Akibo, T.E.; Odukoya, K.E.; Ajiboye, O.O.; Abiona, J.A.; Williams, T.J.; Oke, O.E.; Osinowo, O.A. Growth, Mineral Deposition, and Physiological Responses of Broiler Chickens Offered Honey in Drinking Water during Hot-Dry Season. *Int. J. Zool.* **2012**, *2012*, 1–6. [CrossRef]
113. Adekunle, M.O.; Abioja, M.O.; Abiona, J.A.; Jegede, A.V.; Sodipe, O.G. Rectal Temperature, Heart Rate, Packed Cell Volume and Differential White Blood Cell Count of Laying Pullets to Honey Supplemented Water during Hot–Dry Season. *Slovak J. Anim. Sci.* **2017**, *50*, 15–20.
114. Ali, A.M.; Kunugi, H. Apitherapy for Age-Related Skeletal Muscle Dysfunction (Sarcopenia): A Review on the Effects of Royal Jelly, Propolis, and Bee Pollen. *Foods* **2020**, *9*, 1362. [CrossRef] [PubMed]
115. Collazo, N.; Carpena, M.; Nuñez-Estevez, B.; Otero, P.; Simal-Gandara, J.; Prieto, M.A. Health Promoting Properties of Bee Royal Jelly: Food of the Queens. *Nutrients* **2021**, *13*, 543. [CrossRef] [PubMed]
116. Kurek-Górecka, A.; Górecki, M.; Rzepecka-Stojko, A.; Balwierz, R.; Stojko, J. Bee Products in Dermatology and Skin Care. *Molecules* **2020**, *25*, 556. [CrossRef] [PubMed]

117. Bouamama, S.; Merzouk, H.; Latrech, H.; Charif, N. Royal jelly alleviates the detrimental effects of aging on immune functions by enhancing the in vitro cellular proliferation, cytokines, and nitric oxide release in aged human PBMCS. *J. Food Biochem.* **2021**, *45*, e13619. [CrossRef]

118. Martinello, M.; Mutinelli, F. Antioxidant Activity in Bee Products: A Review. *Antioxidants* **2021**, *10*, 71. [CrossRef]

119. Addeo, N.; Roncarati, A.; Secci, G.; Parisi, G.; Piccolo, G.; Ariano, A.; Scivicco, M.; Rippa, A.; Bovera, F. Potential use of a queen bee larvae meal (Apis mellifera ligustica Spin.) in animal nutrition: A nutritional and chemical-toxicological evaluation. *J. Insects Food Feed* **2021**, *7*, 173–186. [CrossRef]

120. Rahnama, G.; Deldar, H.; Pirsaraei, Z.A.; Kazemifard, M. Oral administration of royal jelly may improve the preservation of rooster spermatozoa. *J. Anim. Physiol. Anim. Nutr.* **2020**, *104*, 1768–1777. [CrossRef]

121. El-Seedi, H.; El-Wahed, A.A.; Yosri, N.; Musharraf, S.G.; Chen, L.; Moustafa, M.; Zou, X.; Al-Mousawi, S.; Guo, Z.; Khatib, A.; et al. Antimicrobial Properties of *Apis mellifera*'s Bee Venom. *Toxins* **2020**, *12*, 451. [CrossRef]

122. Carpena, M.; Nuñez-Estevez, B.; Soria-Lopez, A.; Simal-Gandara, J. Bee Venom: An Updating Review of Its Bioactive Molecules and Its Health Applications. *Nutrients* **2020**, *12*, 3360. [CrossRef]

123. Gu, H.; Han, S.M.; Park, K.-K. Therapeutic Effects of Apamin as a Bee Venom Component for Non-Neoplastic Disease. *Toxins* **2020**, *12*, 195. [CrossRef] [PubMed]

124. Elkomy, A.; El-Hanoun, A.; Abdella, M.; El-Sabrout, K. Improving the reproductive, immunity and health status of rabbit does using honey bee venom. *J. Anim. Physiol. Anim. Nutr.* **2021**. [CrossRef] [PubMed]

125. El-Hanoun, A.; El-Komy, A.; El-Sabrout, K.; Abdella, M. Effect of bee venom on reproductive performance and immune response of male rabbits. *Physiol. Behav.* **2020**, *223*, 112987. [CrossRef]

126. Bakchiche, B.; Temizer, İ.K.; Güder, A.; Çelemli, Ö.G.; Yegin, S.Ç.; Bardaweel, S.K.; Ghareeb, M.A. Chemical Composition and Biological Activities of Honeybee Products From Algeria. *J. Appl. Biotechnol. Rep.* **2020**, *7*, 93–103. [CrossRef]

127. Demir, Z.; Kaya, H. Effect of Bee Pollen Supplemented Diet on Performance, Egg Quality Traits and some Serum Parameters of Laying Hens. *Pak. J. Zool.* **2020**, *52*. [CrossRef]

128. Ali, A.; Kunugi, H. Propolis, Bee Honey, and Their Components Protect against Coronavirus Disease 2019 (COVID-19): A Review of In Silico, In Vitro, and Clinical Studies. *Molecules* **2021**, *26*, 1232. [CrossRef]

129. Abdelnour, S.A.; El-Hack, M.E.A.; Alagawany, M.; Farag, M.R.; ElNesr, S.S. Beneficial impacts of bee pollen in animal production, reproduction and health. *J. Anim. Physiol. Anim. Nutr.* **2018**, *103*, 477–484. [CrossRef]

130. Sanchez, R.D.V.; Ibarra-Arias, F.J.; Torres-Martínez, B.D.M.; Sánchez-Escalante, A.; Torrescano-Urrutia, G.R. Use of natural ingredients in Japanese quail diet and their effect on carcass and meat quality—A review. *Asian-Australas. J. Anim. Sci.* **2019**, *32*, 1641–1656. [CrossRef]

131. Haščík, P.; Trembecká, L.; Bobko, M.; Čuboň, J.; Kačániová, M.; Tkáčová, J. Amino acid profile of broiler chickens meat fed diets supplemented with bee pollen and propolis. *J. Apic. Res.* **2016**, *55*, 324–334. [CrossRef]

132. Trembecká, L.; Haščík, P.; Čuboň, J.; Bobko, M.; Cviková, P.; Hleba, L. Chemical and Sensory Characteristics Of Chicken Breast Meat After Dietary Supplementation With Probiotic Given In Combination With Bee Pollen And Propolis. *J. Microbiol. Biotechnol. Food Sci.* **2017**, *7*, 275–280. [CrossRef]

133. Prakatur, I.; Miškulin, I.; Senčić, Đ.; Pavić, M.; Miškulin, M.; Samac, D.; Galović, D.; Domaćinović, M. The influence of propolis and bee pollen on chicken meat quality. *Vet. Arh.* **2020**, *90*, 617–625. [CrossRef]

134. Ketkar, S.; Rathore, A.S.; Kandhare, A.; Lohidasan, S.; Bochankar, S.; Paradkar, A.; Mahadik, K. Alleviating exercise-induced muscular stress using neat and processed bee pollen: Oxidative markers, mitochondrial enzymes, and myostatin expression in rats. *Integr. Med. Res.* **2015**, *4*, 147–160. [CrossRef] [PubMed]

135. Mohdaly, A.A.; Mahmoud, A.A.; Roby, M.H.; Smetanska, I.; Ramadan, M.F. Phenolic Extract from Propolis and Bee Pollen: Composition, Antioxidant and Antibacterial Activities. *J. Food Biochem.* **2015**, *39*, 538–547. [CrossRef]

136. Khafaji, S.S.O.; Aljanabi, T.K.; Suhailaltaie, S.M. Evaluation the Impact of Different Levels of Propolis on Some Reproductive features in Iraqi Local Roosters. *Adv. Anim. Vet. Sci.* **2018**, *7*. [CrossRef]

137. Viuda-Martos, M.; Ruiz-Navajas, Y.; Fernández-López, J.; Pérez-Alvarez, J.A. Functional Properties of Honey, Propolis, and Royal Jelly. *J. Food Sci.* **2008**, *73*, R117–R124. [CrossRef]

138. Pasupuleti, V.R.; Sammugam, L.; Ramesh, N.; Gan, S.H. Honey, Propolis, and Royal Jelly: A Comprehensive Review of Their Biological Actions and Health Benefits. *Oxidative Med. Cell. Longev.* **2017**, *2017*, 1–21. [CrossRef] [PubMed]

139. Ahangari, Z.; Naseri, M.; Vatandoost, F. Propolis: Chemical Composition and Its Applications in Endodontics. *Iran. Endod. J.* **2018**, *13*, 285–292. [CrossRef] [PubMed]

140. Zweil, H.S.; Zahran, S.M.; El Rahman, M.A.; Desoky, W.M.; Abu Hafsa, S.H.; Mokhtar, A. Effect of Using Bee Propolis as Natural Supplement on Productive and Physiological Performance of Japanese Quail. *Egypt. Poult. Sci. J.* **2016**, *36*, 161–175. [CrossRef]

141. Mehaisen, G.M.K.; Ibrahim, R.M.; Desoky, A.A.; Safaa, H.; El-Sayed, O.A.; Abass, A.O. The importance of propolis in alleviating the negative physiological effects of heat stress in quail chicks. *PLoS ONE* **2017**, *12*, e0186907. [CrossRef] [PubMed]

142. Righi, F.; Pitino, R.; Manuelian, C.; Simoni, M.; Quarantelli, A.; De Marchi, M.; Tsiplakou, E. Plant Feed Additives as Natural Alternatives to the Use of Synthetic Antioxidant Vitamins on Poultry Performances, Health, and Oxidative Status: A Review of the Literature in the Last 20 Years. *Antioxidants* **2021**, *10*, 659. [CrossRef]

143. Pieroni, C.A.; De Oliveira, M.C.; Dos Santos, W.L.R.; Mascarenhas, L.B.; Oliveira, M.A.D. Effect of green propolis on the productivity, nutrient utilisation, and intestinal morphology of Japanese laying quail. *Rev. Bras. Zootec.* **2020**, *49*. [CrossRef]

144. Kostadinović, L.; Lević, J. Effects of Phytoadditives in Poultry and Pigs Diseases. *J. Agron. Technol. Eng. Manag.* **2018**, *1*, 1–7.
145. El-Hack, M.E.A.; Abdelnour, S.A.; Taha, A.E.; Khafaga, A.; Arif, M.; Ayasan, T.; Swelum, A.A.; Abukhalil, M.H.; Alkahtani, S.; Aleya, L.; et al. Herbs as thermoregulatory agents in poultry: An overview. *Sci. Total Environ.* **2020**, *703*, 134399. [CrossRef]
146. Alagawany, M.; Nasr, M.; Al-Abdullatif, A.; Alhotan, R.A.; Azzam, M.M.; Reda, F.M. Impact of dietary cold-pressed chia oil on growth, blood chemistry, haematology, immunity and antioxidant status of growing Japanese quail. *Ital. J. Anim. Sci.* **2020**, *19*, 896–904. [CrossRef]
147. Mishra, N. (Ed.) *Ethnopharmacological Investigation of Indian Spices: Potential Health Benefits of Fenugreek with Multiple Pharmacological Properties*; Advances in Medical Diagnosis, Treatment, and Care; IGI Global: Hershey, PA, USA, 2020; ISBN 978-1-79982-524-1.
148. Tewari, D.; Jóźwik, A.; Łysek-Gładysińska, M.; Grzybek, W.; Adamus-Białek, W.; Bicki, J.; Strzałkowska, N.; Kamińska, A.; Horbańczuk, O.K.; Atanasov, A.G. Fenugreek (*Trigonella foenum-graecum* L.) Seeds Dietary Supplementation Regulates Liver Antioxidant Defense Systems in Aging Mice. *Nutrients* **2020**, *12*, 2552. [CrossRef] [PubMed]
149. Rashid, F.; Bao, Y.; Ahmed, Z.; Huang, J.-Y. Effect of high voltage atmospheric cold plasma on extraction of fenugreek galactomannan and its physicochemical properties. *Food Res. Int.* **2020**, *138*, 109776. [CrossRef]
150. Jain, D.; Bains, K.; Singla, N. Mode of Action of Anti-diabetic Phyto-Compounds Present in Traditional Indian Plants: A Review. *Curr. J. Appl. Sci. Technol.* **2020**, 19–38. [CrossRef]
151. Omri, B.; Manel, B.L.; Jihed, Z.; Durazzo, A.; Lucarini, M.; Romano, R.; Santini, A.; Abdouli, H. Effect of a Combination of Fenugreek Seeds, Linseeds, Garlic and Copper Sulfate on Laying Hens Performances, Egg Physical and Chemical Qualities. *Foods* **2019**, *8*, 311. [CrossRef] [PubMed]
152. Abdouli, H.; Haj-Ayed, M.; Belhouane, S.; Emna, E.H. Effect of Feeding Hens with Fenugreek Seeds on Laying Performance, Egg Quality Characteristics, Serum and Egg Yolk Cholesterol. *J. New Sci.* **2014**, *3*, 1–9.
153. Abbas, R.J. Effect of Using Fenugreek, Parsley and Sweet Basil Seeds as Feed Additives on the Performance of Broiler Chickens. *Int. J. Poult. Sci.* **2010**, *9*, 278–282. [CrossRef]
154. Zheng, Y.; Zhang, Q.; Hu, X. A comprehensive review of ethnopharmacological uses, phytochemistry, biological activities, and future prospects of Nigella glandulifera. *Med. Chem. Res.* **2020**, *29*, 1168–1186. [CrossRef]
155. Williamson, E.M.; Liu, X.; Izzo, A.A. Trends in use, pharmacology, and clinical applications of emerging herbal nutraceuticals. *Br. J. Pharmacol.* **2019**, *177*, 1227–1240. [CrossRef]
156. Aispuro, J.A.M.; Velasco, J.L.F.; Sánchez-Torres, M.T.; Mora, J.L.C. Unconventional plants as a source of phytochemicals for broiler chicken. *AGROProductividad* **2020**, *13*. [CrossRef]
157. Heidary, M.; Hassanabadi, A.; Mohebalian, H. Effects of in Ovo Injection of Nanocurcumin and Vitamin E on Antioxidant Status, Immune Responses, Intestinal Morphology and Growth Performance of Broiler Chickens Exposed to Heat Stress. *J. Livest. Sci. Technol.* **2020**, *8*, 17–27. [CrossRef]
158. Rahman, M.; Kim, S.-J. Effects of dietaryNigella sativaseed supplementation on broiler productive performance, oxidative status and qualitative characteristics of thighs meat. *Ital. J. Anim. Sci.* **2016**, *15*, 241–247. [CrossRef]
159. Boka, J.; Mahdavi, A.H.; Samie, A.H.; Jahanian, R. Effect of different levels of black cumin (*Nigella sativa* L.) on performance, intestinal Escherichia coli colonization and jejunal morphology in laying hens. *J. Anim. Physiol. Anim. Nutr.* **2013**, *98*, 373–383. [CrossRef] [PubMed]
160. Aydin, R.; Karaman, M.; Cicek, T.; Yardibi, H. Black Cumin (*Nigella sativa* L.) Supplementation into the Diet of the Laying Hen Positively Influences Egg Yield Parameters, Shell Quality, and Decreases Egg Cholesterol. *Poult. Sci.* **2008**, *87*, 2590–2595. [CrossRef] [PubMed]
161. Zhang, M.; Zhao, R.; Wang, D.; Wang, L.; Zhang, Q.; Wei, S.; Lu, F.; Peng, W.; Wu, C. Ginger (*Zingiber officinale* Rosc.) and its bioactive components are potential resources for health beneficial agents. *Phytother. Res.* **2020**, *35*, 711–742. [CrossRef]
162. Ma, R.-H.; Ni, Z.-J.; Zhu, Y.-Y.; Thakur, K.; Zhang, F.; Zhang, Y.-Y.; Hu, F.; Zhang, J.-G.; Wei, Z.-J. A recent update on the multifaceted health benefits associated with ginger and its bioactive components. *Food Funct.* **2020**, *12*, 519–542. [CrossRef] [PubMed]
163. Herve, T.; Raphaël, K.J.; Ferdinand, N.; Herman, N.V.; Marvel, N.M.W.; D'Alex, T.C.; Vitrice, F.T.L. Effects of Ginger (*Zingiber officinale*, Roscoe) Essential Oil on Growth and Laying Performances, Serum Metabolites, and Egg Yolk Antioxidant and Cholesterol Status in Laying Japanese Quail. *J. Vet. Med.* **2019**, *2019*, 1–8. [CrossRef]
164. El-Hack, M.E.A.; Alagawany, M.; Shaheen, H.; Samak, D.; Othman, S.I.; Allam, A.A.; Taha, A.E.; Khafaga, A.F.; Arif, M.; Osman, A.; et al. Ginger and Its Derivatives as Promising Alternatives to Antibiotics in Poultry Feed. *Animals* **2020**, *10*, 452. [CrossRef]
165. Oleforuh-Okoleh, V.U.; Ndofor-Foleng, H.M.; Olorunleke, S.O.; Uguru, J.O. Evaluation of Growth Performance, Haematological and Serum Biochemical Response of Broiler Chickens to Aqueous Extract of Ginger and Garlic. *J. Agric. Sci.* **2015**, *7*. [CrossRef]
166. He, F.-Y.; Kim, H.-W.; Hwang, K.-E.; Song, D.-H.; Kim, Y.-J.; Ham, Y.-K.; Kim, S.-Y.; Yeo, I.-J.; Jung, T.-J.; Kim, C.-J. Effect of Ginger Extract and Citric Acid on the Tenderness of Duck Breast Muscles. *Food Sci. Anim. Resour.* **2015**, *35*, 721–730. [CrossRef]
167. Ajitomi, A.; Inoue, Y.; Horita, M.; Nakaho, K. Bacterial wilt of three Curcuma species, C. longa (turmeric), C. aromatica (wild turmeric) and C. zedoaria (zedoary) caused by Ralstonia solanacearum in Japan. *J. Gen. Plant Pathol.* **2015**, *81*, 315–319. [CrossRef]
168. Mythri, R.B. Curcumin: A Potential Neuroprotective Agent in Parkinson's Disease. *Curr. Pharm. Des.* **2012**, *18*, 91–99. [CrossRef] [PubMed]

169. Akhavan-Salamat, H.; Ghasemi, H.A. Alleviation of chronic heat stress in broilers by dietary supplementation of betaine and turmeric rhizome powder: Dynamics of performance, leukocyte profile, humoral immunity, and antioxidant status. *Trop. Anim. Health Prod.* **2015**, *48*, 181–188. [CrossRef]

170. Aggarwal, B.B.; Sundaram, C.; Malani, N.; Ichikawa, H. Curcumin: The Indian solid gold BT. In *The Molecular Targets and Therapeutic Uses of Curcumin in Health and Disease*; Aggarwal, B.B., Surh, Y.-J., Shishodia, S., Eds.; Advances in Experimental Medicine and Biology; Springer US: Boston, MA, USA, 2007; pp. 1–75. ISBN 978-0-387-46401-5.

171. Priyadarsini, K. Free Radical Reactions of Curcumin in Membrane Models. *Free Radic. Biol. Med.* **1997**, *23*, 838–843. [CrossRef]

172. Swathi, B.; Gupta, P.; Nagalakshmi, D.; Raju, M. Efficacy of turmeric (*Curcuma longa*) as antioxidant in combating heat stress in broiler chicken. *Indian J. Poult. Sci.* **2016**, *51*, 48. [CrossRef]

173. Sadeghi, A.; Moghaddam, M. The Effects of Turmeric, Cinnamon, Ginger and Garlic Powder Nutrition on Antioxidant Enzymes' Status and Hormones Involved in Energy Metabolism of Broilers during Heat Stress. *Iran. J. Appl. Anim. Sci.* **2018**, *8*, 125–130.

174. El-Maaty, A.; Hayam, M.; Rabie, M.; El-Khateeb, A.; El-Maaty, A.A.; El-Khateeb, A. Response of Heat-Stressed Broiler Chicks to Dietary Supplementation with Some Commercial Herbs. *Asian J. Anim. Vet. Adv.* **2014**, *9*, 743–755. [CrossRef]

175. Alagawany, M.; Farag, M.R.; Dhama, K. Nutritional and Biological Effects of Turmeric (*Curcuma longa*) Supplementation on Performance, Serum Biochemical Parameters and Oxidative Status of Broiler Chicks Exposed to Endosulfan in the Diets. *Asian J. Anim. Vet. Adv.* **2015**, *10*, 86–96. [CrossRef]

176. Swamy, M.K.; Sinniah, U.R. A Comprehensive Review on the Phytochemical Constituents and Pharmacological Activities of Pogostemon cablin Benth.: An Aromatic Medicinal Plant of Industrial Importance. *Molecules* **2015**, *20*, 8521–8547. [CrossRef] [PubMed]

177. Lee, S.-J.; Umano, K.; Shibamoto, T.; Lee, K.-G. Identification of volatile components in basil (*Ocimum basilicum* L.) and thyme leaves (*Thymus vulgaris* L.) and their antioxidant properties. *Food Chem.* **2005**, *91*, 131–137. [CrossRef]

178. Abdallah, A.; Zhang, P.; Zhong, Q.; Sun, Z. Application of Traditional Chinese Herbal Medicine By-products as Dietary Feed Supplements and Antibiotic Replacements in Animal Production. *Curr. Drug Metab.* **2019**, *20*, 54–64. [CrossRef] [PubMed]

179. Calín-Sánchez, K.; Figiel, A.; Lech, K.; Szumny, A.; Carbonell-Barrachina, A.; Carbonell-Barrachina, A. Effects of Drying Methods on the Composition of Thyme (*Thymus vulgaris* L.) Essential Oil. *Dry. Technol.* **2013**, *31*, 224–235. [CrossRef]

180. Aziz, M.; Karboune, S. Natural antimicrobial/antioxidant agents in meat and poultry products as well as fruits and vegetables: A review. *Crit. Rev. Food Sci. Nutr.* **2016**, *58*, 1–26. [CrossRef] [PubMed]

181. Ding, X.; Yu, Y.; Su, Z.; Zhang, K. Effects of essential oils on performance, egg quality, nutrient digestibility and yolk fatty acid profile in laying hens. *Anim. Nutr.* **2017**, *3*, 127–131. [CrossRef]

182. Yalçin, S.; Eser, H.; Onbaşilar, İ.; Yalçin, S. Effects of dried thyme (*Thymus vulgaris* L.) leaves on performance, some egg quality traits and immunity in laying hens. *Ank. Üniversitesi Vet. Fakültesi Derg.* **2020**, *67*, 303–311. [CrossRef]

183. Hassan, F.; Awad, A. Impact of thyme powder (*Thymus vulgaris* L.) supplementation on gene expression profiles of cytokines and economic efficiency of broiler diets. *Environ. Sci. Pollut. Res.* **2017**, *24*, 15816–15826. [CrossRef] [PubMed]

184. El-Ashram, S.; Abdelhafez, G.A. Effects of phytogenic supplementation on productive performance of broiler chickens. *J. Appl. Poult. Res.* **2020**, *29*, 852–862. [CrossRef]

185. Attia, Y.A.; Bakhashwain, A.A.; Bertu, N.K. Thyme oil (*Thyme vulgaris* L.) as a natural growth promoter for broiler chickens reared under hot climate. *Ital. J. Anim. Sci.* **2016**, *16*, 275–282. [CrossRef]

186. Abdulkarimi, R.; Daneshyar, M.; Aghazadeh, A. Thyme (*Thymus vulgaris*) extract consumption darkens liver, lowers blood cholesterol, proportional liver and abdominal fat weights in broiler chickens. *Ital. J. Anim. Sci.* **2011**, *10*, e20. [CrossRef]

187. Haselmeyer, A.; Zentek, J.; Chizzola, R. Effects of thyme as a feed additive in broiler chickens on thymol in gut contents, blood plasma, liver and muscle. *J. Sci. Food Agric.* **2014**, *95*, 504–508. [CrossRef]

188. Hashemipour, H.; Kermanshahi, H.; Golian, A.; Veldkamp, T. Effect of thymol and carvacrol feed supplementation on performance, antioxidant enzyme activities, fatty acid composition, digestive enzyme activities, and immune response in broiler chickens. *Poult. Sci.* **2013**, *92*, 2059–2069. [CrossRef] [PubMed]

189. Pournazari, M.; Islamic Azad University; Qotbi, A.A.; Seidavi, A.; Corazzin, M.; Di Udine, U. Prebiotics, probiotics and thyme (*Thymus vulgaris*) for broilers: Performance, carcass traits and blood variables. *Rev. Colomb. Cienc. Pecu.* **2017**, *30*, 3–10. [CrossRef]

190. Adefegha, S.A.; Oyeleye, S.I.; Akintemi, A.; Okeke, B.M.; Oboh, G. Thyme (Thymus vulgaris) leaf extract modulates purinergic and cholinergic enzyme activities in the brain homogenate of 5-fluorouracil administered rats. *Drug Chem. Toxicol.* **2019**, *43*, 43–50. [CrossRef]

191. Surai, P.F.; Kochish, I.I.; Fisinin, V.I.; Kidd, M.T. Antioxidant Defence Systems and Oxidative Stress in Poultry Biology: An Update. *Antioxidants* **2019**, *8*, 235. [CrossRef]

192. Placha, I.; Ocelova, V.; Chizzola, R.; Battelli, G.; Gai, F.; Bacova, K.; Faix, S. Effect of thymol on the broiler chicken antioxidative defence system after sustained dietary thyme oil application. *Br. Poult. Sci.* **2019**, *60*, 589–596. [CrossRef] [PubMed]

193. El-Ghousein, S.S.; Al-Beitawi, N.A. The Effect of Feeding of Crushed Thyme (*Thymus valgaris* L.) on Growth, Blood Constituents, Gastrointestinal Tract and Carcass Characteristics of Broiler Chickens. *J. Poult. Sci.* **2009**, *46*, 100–104. [CrossRef]

194. Hernández, F.; Madrid, J.; García, V.; Orengo, J.; Megias, M. Influence of two plant extracts on broilers performance, digestibility, and digestive organ size. *Poult. Sci.* **2004**, *83*, 169–174. [CrossRef] [PubMed]

195. Nouri, A. Chitosan nano-encapsulation improves the effects of mint, thyme, and cinnamon essential oils in broiler chickens. *Br. Poult. Sci.* **2019**, *60*, 530–538. [CrossRef] [PubMed]

 MDPI

Article

Replacing Maize Grain with Ancient Wheat Lines By-Products in Organic Laying Hens' Diet Affects Intestinal Morphology and Enzymatic Activity

Nicola Francesco Addeo [1], Basilio Randazzo [2], Ike Olivotto [2], Maria Messina [3], Francesca Tulli [3], Nadia Musco [1,*], Giovanni Piccolo [1], Antonino Nizza [4], Carmelo Di Meo [1] and Fulvia Bovera [1]

[1] Department of Veterinary Medicine and Animal Production, University of Napoli Federico II, Via Federico Delpino 1, 80137 Napoli, Italy; nicolafrancesco.addeo@unina.it (N.F.A.); giovanni.piccolo@unina.it (G.P.); carmelo.dimeo@unina.it (C.D.M.); bovera@unina.it (F.B.)

[2] Department of Life and Environmental Sciences, Marche Polytechnic University, Via Brecce Bianche, 60131 Ancona, Italy; b.randazzo.live@gmail.com (B.R.); i.olivotto@staff.univpm.it (I.O.)

[3] Department of AgriFood, Environment and Animal Science, University of Udine, Via Sondrio, 2, 33100 Udine, Italy; maria.messina@uniud.it (M.M.); francesca.tulli@uniud.it (F.T.)

[4] Department of Agricultural Sciences, University of Naples Federico II, 80055 Portici, Italy; nizza@unina.it

* Correspondence: nadia.musco@unina.it

Abstract: The effects of replacement of maize grain with ancient wheat by-products on intestinal morphometry and enzymatic activity in laying hens was studied. Eighty hens were divided into two groups (40 each, 8 replicates, 5 hens/replicate) fed two isoproteic and isoenergetic diets. In the treated group, part of the maize was replaced by a mix of ancient grains (AGs) middling, in a 50:50 ratio of *Triticum aestivum* L. var. spelta (spelt) and *Triticum durum dicoccum* L. (emmer wheat). The AG diet affected the weight of all the large intestine tracts, decreasing the weight of caeca ($p < 0.01$) and increasing those of colon ($p < 0.01$), rectum and cloaca ($p < 0.05$). Villus height in the AG group was higher ($p < 0.01$) than the control for the duodenum and jejunum, while for the ileum, the control group showed the highest values ($p < 0.01$). The submucosa thickness was higher ($p < 0.01$) in the control group for the duodenum and ileum, while the jejunum for the AG group showed the highest ($p < 0.05$) submucosa thickness. The crypts depth was higher ($p < 0.01$) in the control group for the duodenum and ileum. Enzyme activity was enhanced by AGs ($p < 0.01$) in the duodenum. Regarding the jejunum, sucrase-isomaltase and alkaline phosphatase had higher activity ($p < 0.05$ and $p < 0.01$, respectively) in the AG group. In the ileum, sucrase-isomaltase showed higher activity ($p < 0.01$) in the control group, while alkaline phosphatase showed the highest values ($p < 0.05$) in the AG group. Overall, results suggested that the dietary inclusion of AGs exerted positive effects in hens, showing an improved intestinal function.

Keywords: ancient grains; organic farm; intestinal morphometry; enzymatic activity; animal performance

Citation: Addeo, N.F.; Randazzo, B.; Olivotto, I.; Messina, M.; Tulli, F.; Musco, N.; Piccolo, G.; Nizza, A.; Di Meo, C.; Bovera, F. Replacing Maize Grain with Ancient Wheat Lines By-Products in Organic Laying Hens' Diet Affects Intestinal Morphology and Enzymatic Activity. *Sustainability* **2021**, *13*, 6554. https://doi.org/10.3390/su13126554

Academic Editors: Nikola Puvača, Vincenzo Tufarelli and Eva Voslarova

Received: 6 May 2021
Accepted: 5 June 2021
Published: 8 June 2021

Publisher's Note: MDPI stays neutral with regard to jurisdictional claims in published maps and institutional affiliations.

1. Introduction

The term "ancient wheat" indicates primitive grains that never underwent selection or breeding, thus retaining their wild ancestors' patterns, such as high individual variety, brittle rachis, and low harvest index [1]. In the last decade, a rediscovery of ancient varieties took place, aiming to produce high value food products with great health benefits [2,3]. These beneficial properties were attributed to the presence of some nutrients, especially unsaturated fatty acids, soluble fibers, minerals, vitamins, and phytochemicals [3–7]. The highest concentration of such nutrients occurs in the outer layers of grains [8,9], thus explaining why the reduced risk of developing several diseases has been associated with an increased consumption of whole grains [10–12]. After milling, only the endosperm of the whole wheat grain is used to make white flour, whereas the bran and middling are used as by-products [13], which could also represent an important added value brought to animal feeding. Although durum wheat represents the vast world production of wheat, with the main production and cultivation areas concentrated in the Mediterranean [14], other cultivars such as spelt einkorn (*Triticum monococcum* L.), spelt ("ancient grains") are still produced in small quantities (mainly for traditional foods) in recent years to meet the growing interest of the natural food market.

The compelling argument in favor of ancient wheats is their environmentally friendly production—in fact, growth with few agronomic practices, environmental sustainability, and possible use in marginal lands or under organic growing conditions—thus, they can be produced in a more sustainable way with few external inputs [15,16].

This study is the completion of a previous trial [17] where laying performance, serum biochemistry, and the physical quality of the eggs of hens fed with ancient wheats (spelt and emmer wheat) were investigated. The objective of the present trial, performed in an organic laying hen farm, was to evaluate the effects of a partial replacement of maize grain with local ancient wheats by-products on the intestinal morphometry and brush border enzymatic activity of 36-week-old laying hens.

2. Materials and Methods

2.1. Animals

The animals used in this study were treated following the Directive 63/2010/EEC regarding animal welfare and the safeguard of experimental animals. This research was approved by the Ethical Animal Care and Use Committee of the University of Naples Federico II, Italy (number 2017/0017676). Experiments were performed in an organic laying hens farm in Avellino (Italy) for 14 weeks (February–May 2019).

Eighty Hy-Line W-36 Single Comb White Leghorn hens, aged 18 weeks and with an average body weight of 1.57 ± 0.09 kg, were randomly divided into two groups (40 animals each; 8 replicates of 5 hens each/group). Each group was stabled in a free-range area equipped with an indoor recover for the night. The available space for each hen was 0.45 m^2 indoor and 5 m^2 in the outdoor areas. A detailed description of the hens' housing system was presented in a previous paper [17].

2.2. Diets

Every morning both groups were fed two isoproteic and isoenergetic diets; the differences in ingredients are reported in Table 1.

Table 1. Ingredients and chemical composition of the diets used in the trial.

Ingredients (%)	Control Diet	Ancient Grains Diet
Maize grain (*Zea mays*)	59.3	25.0
Emmer wheat middling (*Triticum durum dicoccum*)	-	18.15
Common wheat middling (*Triticum aestivum*)	-	18.15
Soybean meal	30.0	27.0
Salt	0.2	0.2
Calcium carbonate	8.0	8.0
Vegetable oil	1.0	2.0
Vitamin–mineral premix	1.0	1.0
Monocalcium Phosphate	0.5	0.5
Chemical characteristics		
Dry matter (DM) [1], %	88.4	87.7
Crude protein [1], % DM	19.5	20.2
Ether extract [1], % DM	4.12	4.79

Table 1. *Cont.*

Ingredients (%)		
	Control Diet	**Ancient Grains Diet**
Neutral detergent fiber [1], % DM	10.50	13.80
Acid detergent fiber [1], % DM	5.61	5.83
Acid detergent lignin [1], % DM	0.80	0.91
Ca [2], % DM	3.87	3.89
P [2], % DM	0.52	0.54
Methionine [2], % DM	0.57	0.68
Lysine [2], % DM	1.09	1.31
ME [2], Kcal/kg DM	3.09	3.15

Vitamin–mineral premix contained the following per kg: retynil acetate 10,000 IU, Vit. D3 3000 IU, Vit. E 45 mg, Vit. B6 4.0 mg, Vit. B12 0.02 mg, Vit. K3 3.5 mg, d-pantothenate calcium 13.9 mg, niacin 50 mg, biotin 0.2 mg, ferrous sulfate 122 mg, cupric sulfate 96 mg, zinc oxide 124 mg, manganese oxide 129 mg, anhydrous calcium iodate 1.5 mg, sodium selenite 0.44 mg.[1] determined values; [2]: calculated values.

The control group (C) received a standard diet containing organic maize and soybean meal and formulated to exceed the requirements reported in the *W-36 Commercial Layers Management Guide* (Hi-line, 2016); in the ancient grains (AG) group around 57.8% of maize grain was substituted with a mix of ancient grains middling in a 50:50 ratio of *Triticum aestivum* L. var. spelta (spelt) and *Triticum durum dicoccum* L. (emmer wheat). The ingredients, ground to the same particle size to avoid a possible influence on the animal feed choice, were mixed in a small local mill; the same mill produced the millings used in the trial. The analysis of the diets (chemical and nutritional) was performed following the AOAC indications [18] (ID number: 2001.12, 978.04, 920.39, 978.10, and 930.05 for dry matter (DM), crude protein (CP), ether extract (EE), crude fiber (CF), and ash, respectively). The methods of Van Soest et al. [19] were used to assay neutral detergent fiber (NDF), acid detergent fiber (ADF), and acid detergent lignin (ADL). The diets' metabolizable energy was calculated according to the NRC [20]. Ca, P, methionine, and lysine contents were estimated according to the diet ingredients content [15,21]. The chemical–nutritional characteristics of the grains are depicted in Table 2.

Table 2. Chemical characteristics of the cereal sources used in the trial (% as feed basis).

Chemical Characteristics	Maize Grain	Spelt Wheat Middlings	Emmer Wheat Middlings
Dry matter [1] %	88.7	87.9	89.2
Ash [1] %	1.3	3.4	2.8
Crude protein [1] %	7.65	13.2	11.8
Ether extract [1] %	3.7	3.5	2.7
Neutral detergent fiber [1] %	8.9	21.7	13.2
Acid detergent fiber [1] %	3.2	5.4	4.3
Acid detergent lignin [1] %	0.9	1.4	1.1
Ca [2] %	0.02	0.04	0.04
P [2] %	0.26	0.47	0.51
ME [2], kcal/kg	3.32	2.87	3.43

[1] determined values; [2] calculated values.

At the end of the trial (36 weeks of age), after 12 hours of fasting, 16 hens per group (2 per replicate, 32 hens in total) were weighed and slaughtered in a specialized slaughterhouse, the digestive tract was removed and weighed, the different intestinal tracts were identified, weighed, measured for length, and properly stored for further analysis.

For each group, small intestine tracts from 8 animals were collected for histological analysis. Samples were washed using a pH 7 isotonic ice-cold saline buffer, dried with absorbent paper, and the duodenum, jejunum, and ileum were separated, weighed, and stored in aluminum foil at −20 °C to be later used for the analysis of the brush border membrane (BBM) enzymes.

2.3. Villus and Crypt Morphometry

Samples (0.5 cm) from the duodenum, jejunum, and ileum of 8 animals per group were prepared for histological analysis according to Zarantoniello et al. [22] and Moniello et al. [23]. Briefly, samples were fixed by immersion in 4% phosphate-buffered paraformaldehyde for 48 h. Samples were then washed in phosphate-buffered saline solution (pH = 7.4), dehydrated in graded ethanol solutions, and embedded in paraffin. Cross sections (5 μm) at an interval of 200 μm were stained with Mayer's hematoxylin and eosin (H&E) and Alcian blue (Ab) for the acid mucopolysaccharide-secreting cells (Ab+) detection. Stained sections were examined under a Zeiss Axio Imager A2 microscope according to Zarantoniello et al. [24]. For the evaluation of morphometric parameters (intestinal fold height, submucosa thickness, and crypts depth), 10 random microscopic fields from each section of the duodenum, jejunum, and ileum were acquired by a microscope equipped with a color digital camera Axiocam 503 (Zeiss, Oberkochen, Germany). Data obtained were analyzed by mean of unpaired t test (significance $p < 0.05$) and reported as means ± SD.

2.4. Brush Border Membrane Enzymes Activity

The evaluation of BBM enzymes was done as reported by Shirazy-Beechey et al. [25] with few changes. All steps were done at 4 °C. Briefly, 100 mg of tissue was diluted 1:10 with a buffer (100 mM mannitol, 2 mM Hepes-tris, pH 7.1), with $MgCl_2$ added at a final concentration of 10 mM, and crushed with a TissueLyser (TissueLyser II, Qiagen, Germany) at 30 Hz for 1 min. After a first centrifugation at $2000\times g$ at 4 °C for 10 min, the supernatant was centrifuged at $15,000\times g$ at 4 °C for 10 min. The resulting supernatant was stored at -20 °C until the analysis of maltase, sucrase-isomaltase (SI), L-aminopeptidase (L-ANP), and alkaline phosphatase (IAP) BBM enzyme activity.

The hydrolysis of sucrose and maltose by the mucosal maltase and sucrase was determined as reported by Tibaldi et al. [26].

The intestine alkaline phosphatase (IAP) activity was determined by using a commercial kit (Paramedical, Pontecagnano Faiano, Sa, Italy) using the manufacturer's instructions.

L-aminopeptidase (L-ANP) was determined as reported by Vizcaino et al. [27].

Total proteins were determined according to Bradford [28] (Sigma-Aldrich cat. no. B6916) and bovine serum albumin (Sigma-Aldrich cat. no. 0834) as a standard.

One unit (U) of enzyme activity is the amount of enzyme that transforms or hydrolyzes 1 μmol of the substrate per minute. Specific enzyme activity was calculated as 1 U of the enzyme activity per mg of protein.

2.5. Statistical Analysis

The normal distribution of data and error was evaluated using the Shapiro–Wilk test (SAS, 2002). Data were processed by one-way ANOVA according to the following model:

$$Yij = m + Di + eij$$

where Y is the single observation, m the general mean, D the effect of the diet (i = control or ancient grains), e is the error using the PROC GLM [20]. The comparison between the means was performed by Tukey's test [29]. The results were expressed as average value and the significance level was set at $p \leq 0.05$; p values < 0.10 were considered as a tendency.

3. Results

The relative weights of the whole digestive tract, proventriculus, gizzard, liver, spleen, and abdominal fat for the different dietary treatments are reported in Table 3.

Table 3. Body weight and relative weight of empty gut, proventriculus, gizzard, liver, spleen, and abdominal fat in hens fed the experimental diets over 16 weeks.

	Control Diet	Ancient Grains Diet	RMSE	*p*-Value
Body weight, g	1498.7	1521.2	97.2	0.6441
Empty gut, % BW	8.02	8.84	0.71	0.0725
Gizzard, % BW	1.40	1.50	0.27	0.1106
Proventriculus, % BW	0.51	0.53	0.07	0.0854
Liver, % BW	2.81	2.65	0.21	0.4396
Spleen, % BW	0.14b	0.19a	0.05	0.0316
Abdominal fat, % BW	1.76a	0.63b	0.086	0.0112

RMSE: root mean square error; a, b: $p < 0.05$. $n = 16$.

The spleen weight expressed as BW percentage was higher in the AG group ($p < 0.05$), while the percentage of abdominal fat showed higher values in the control group ($p < 0.05$).

Table 4 shows the relative weights and lengths of the small intestine tracts of the hens, according to the dietary treatments. The duodenum and jejunum lengths exhibited the highest weights in the control group ($p < 0.05$).

Table 4. Relative weight and length (% body weight) of the empty duodenum, jejunum, and ileum in hens fed the experimental diets over 16 weeks.

	Control Diet	Ancient Grains Diet	RMSE	*p*-Value
Duodenum weight	1.40	1.41	0.11	0.8439
Jejunum weight	1.73	1.70	0.83	0.6241
Ileum weight	1.03	1.15	0.09	0.4744
Duodenum length	2.52a	2.16b	0.19	0.0156
Jejunum length	4.70a	4.12b	0.36	0.0372
Ileum length	3.40	3.28	0.35	0.6636

RMSE: root mean square error; a, b: $p < 0.05$. $n = 16$.

Table 5 reports the relative weights and lengths of the large intestine tracts of the hens according to the dietary treatments. Ancient grains affected the weight of all the large intestine tracts, decreasing the relative weight of caeca ($p < 0.01$) and increasing that of colon ($p < 0.01$), rectum and cloaca ($p < 0.05$). Conversely, only the relative length of the caeca was significantly decreased by the treatment ($p < 0.05$). Colon length % tended ($p = 0.0510$) to be higher in the AG group.

Table 5. Relative weight and length (% body weight) of the empty caeca, colon, rectum, and cloaca in hens fed the experimental diets over 16 weeks.

	Control Diet	Ancient Grains Diet	RMSE	*p*-Value
Caeca weight	0.66A	0.61B	0.06	0.0002
Colon weight	0.49B	0.63A	0.05	0.0126
Rectum weight	0.26b	0.31a	0.02	0.0262
Cloaca weight	0.56b	0.64a	0.04	0.0356
Caeca length	2.75a	2.42b	0.24	0.0371
Colon length	1.21	1.40	0.10	0.0510
Rectum length	0.64	0.62	0.04	0.1722
Cloaca length	0.27	0.28	0.05	0.1798

RMSE: root mean square error; A, B: $p < 0.01$; a, b: $p < 0.05$. $n = 16$.

Table 6 shows the morphometric evaluation performed on the duodena, jejuna, and ilea of hens according to the dietary treatment. The villus length in the AG group was higher ($p < 0.01$) than the control in the duodenum and jejunum while in the ileum, the control group showed the highest values of villus length ($p < 0.01$).

Table 6. Morphometric evaluations performed on the duodena, jejuna, and ilea of hens fed the experimental diets over 16 weeks.

	Control Diet	Ancient Grains Diet	RMSE	*p*-Value
	Villus length, μm			
Duodenum	888.1B	1116.5A	103.5	<0.0001
Jejunum	941.8B	1156.8A	108.5	<0.0001
Ileum	953.2A	660.6B	198.9	0.0016
	Submucosa thickness, μm			
Duodenum	64.2A	33.6B	6.63	0.0002
Jejunum	43.22b	51.65a	7.72	0.0138
Ileum	71.28A	45.39B	18.98	0.0030
	Crypts depth, μm			
Duodenum	410.58A	317.0B	45.590	<0.0001
Jejunum	434.9	360.1	115.7	0.1274
Ileum	419.4A	263.2B	10.41	0.0010
	Villus:Crypt ratio			
Duodenum	2.14B	3.52A	0.18	0.0043
Jejunum	2.16B	3.21A	0.20	0.0062
Ileum	2.27b	2.50a	0.16	0.0121

RMSE: root mean square error; A, B: $p < 0.01$; a, b: $p < 0.05$. $n = 8$.

The submucosa thickness was higher ($p < 0.01$) in the control group for the duodenum and ileum, while the jejunum in the AG group showed the highest ($p < 0.05$) submucosa thickness. The crypts depth was higher ($p < 0.01$) in the control group for the duodenum and ileum. The villus:crypt ratio in the three tracts of the small intestine was higher for the AG group in the duodenum and jejunum ($p < 0.01$) and in the ileum tract ($p < 0.05$).

Table 7 shows the specific activity of the brush border enzymes in the small intestine of hens according to the dietary treatment. The activity of all the evaluated enzymes (maltase, sucrase-isomaltase, L-aminopeptidase, and intestinal alkaline phosphatase) was enhanced by the presence of AGs ($p < 0.01$) in the duodenum.

Table 7. Specific activity of intestinal brush border membrane (BBM) enzymes measured in the different digestive tracts of the hens fed the experimental diets over 16 weeks.

Enzymatic Activity	Control Diet	Ancient Grains Diet	RMSE	*p*-Value
Duodenum				
Maltase, U	22.49[B]	46.94[A]	7.82	0.000
SI, U	5.95[B]	13.10[A]	4.29	0.005
L-ANP, U	1.58[B]	3.28[A]	0.67	0.000
IAP, mU	290.83[B]	550.61[A]	131.58	0.001
Jejunum				
Maltase, U	32.97	41.19	17.16	0.339
SI, U	7.43[b]	13.35[a]	4.81	0.023
L-ANP, U	2.44	2.25	0.83	0.656
IAP, mU	397.33[B]	817.57[A]	245.20	0.003
Ileum				
Maltase, U	34.18	32.23	8.84	0.651
SI, U	10.77[A]	6.26[B]	2.29	0.002
L-ANP, U	2.73	2.48	0.544	0.360
IAP, mU	321.16[b]	493.26[a]	138.18	0.032

SI: sucrase-isomaltase; L-ANP: L-aminopeptidase; IAP: intestinal alkaline phosphatase. RMSE: root mean square error; A, B: $p < 0.01$; a, b: $p < 0.05$. $n = 8$.

Regarding the jejunum, SI and IAP had higher activity ($p < 0.05$ and $p < 0.01$, respectively) in the AG group than in the control group. In the ileum, SI showed higher activity ($p < 0.01$) in the control group, while IAP showed the highest values ($p < 0.05$) in the AG group.

Histological analysis performed on the duodenum, jejunum, and ileum exhibited intact intestinal mucosa in both groups in all the analyzed samples, showing a continuous epithelial layer forming the absorptive mucosa and a low number of exfoliated cells in the lumen (Figure 1).

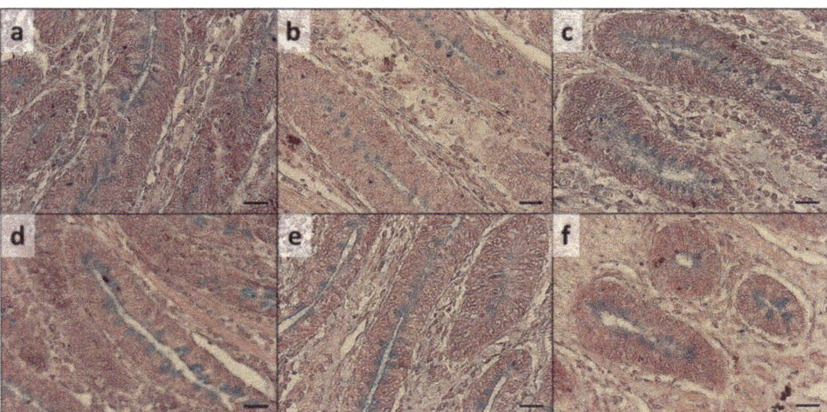

Figure 1. Ab+ mucous cells (arrowheads) in duodenum (**a,d**), jejunum (**b,e**), and ileum (**c,f**) crypts from C (**a–c**) and AG groups (**d–f**) of hens fed the experimental diets over 16 weeks. Alcian blue. Scale = 20 μm.

A regular degree of basal crypts was observed in all the intestinal traits analyzed, with a comparable Ab+ mucous cells distribution between the experimental groups (Figure 2). Nevertheless, melanomacrophage intra-epithelial influx (Figure 2) was observed in 75% of group C duodenum samples, while it was not detected in any of the AG group samples.

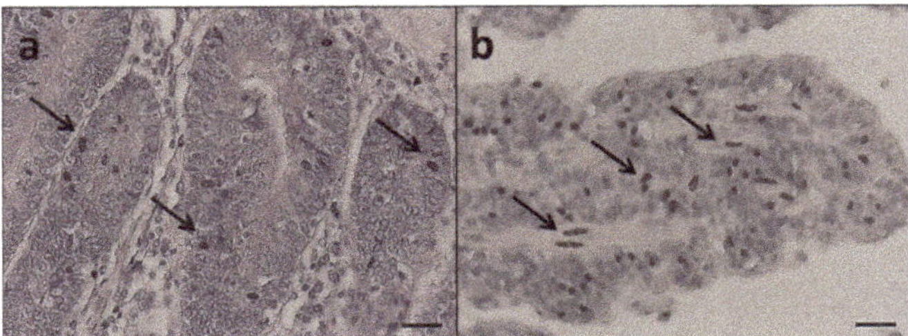

Figure 2. Group C duodenum crypts (**a**) and lamina propria of jejunum villi (**b**) showing melanomacrophage influx (arrows) in hens fed the experimental diets over 16 weeks. H&E. Scale = 20 μm.

4. Discussion

The inclusion of ancient grains in the diet of hens did not affect the weight of the small intestine tracts mainly involved in the digestion and absorption of the nutrients, but the length of both duodenum and jejunum was lower by 13.36 and 11.10% than the control group, respectively. According to the first part of this trial [17], feed intake was lower in the hens of the AG group than in the control one (120.9 vs. 134.1 g/d, respectively, $p < 0.01$), probably due to a low palatability of the ancient wheats compared to the maize [17], while the digestibility of organic matter (75.12 vs. 75.98%), crude protein (86.23 vs. 85.47%) and ether extract (92.48 vs. 91.87%) were unaffected by the dietary treatment. The higher levels in methionine and lysine are due to their higher content in ancient wheats than in maize.

However, the essential amino acid levels were calculated based on literature values and not determined.

The higher feed intake could justify the higher percentage of abdominal fat recorded in hens fed the control diet, in contrast with the results of El-Katcha et al. [30]. These authors reported that including wheat (in ratios of 25 and 50%) instead of corn in the diet of broiler chickens did not significantly reduce abdominal fat weight compared with a broiler chicken group fed a corn–soybean-based diet. However, the previous trial from Lombardi et al. [17] showed that wheat inclusion did not impair intestinal nutrient digestibility and the hen's final weight.

Such a result could be related to the different intestinal morphology between the groups. The morphology of the small intestine is often used to explore its functionality and, generally an increased villus height, as recorded in the present trial in the duodenum and jejunum for the AG group, is indicative of an improved intestinal function [31]. It seems that when ancient grains were administered in the diet, the hens' intestinal tracts increased absorptive tissue rather than length, which would involve a higher energy requirement. The improvement in intestine absorptive surface may be related to the presence of AG peptides and to the presence of insoluble non-polysaccharide starch (NPS) which, different from soluble viscous NSP, is able to enhance intestine welfare [32,33]. In any case, the effect of bioactive peptides from cereal grains has already been recognized in humans [34,35]. Another interesting consideration is that, in chickens, the ileal villi are smaller than those of the previous tracts of the small intestine, and, in hens fed corn–soybean-based diets, very few nutrients are available beyond the jejunum [36–38].

In addition, in the present trial, the highest feed intake recorded in the control group, associated with an unmodified nutrient digestibility in comparison to the AG group [17], could have produced a higher amount of undigested starch particles in the ilea of the hens fed the corn–soybean-based diet, and this could be responsible for the higher height of villi in the ilea of hens of the control group. It is noteworthy that the ileum plays a significant role in digestion and absorption of undigested starch in chickens [39,40].

Also, Yamauchi [41] reported that increasing the load of nutrients from the duodenum to the ileum (for both jejunum dissection and different diets) may stimulate the absorptions in the ileum, thus resulting in a compensatory development of the villi as observed in the control group ilea. It is known that the presence of longer villi is due to the activation of cell mitosis in the crypts [42]; as a consequence, a larger crypt area means an increased cell production. In our trial, only the crypt depth was recorded, being higher in the control group for both duodenum and ileum tracts. In addition, the villus height-to-crypt depth ratio (VH/CD), which is a measure of the epithelial cell turnover [41], was lower in the control group for the three small intestine tracts. Sozcu and Ipek [43] studied the effects of lignocellulose supplementation on jejunal histomorphology of laying hens, including increasing levels (0.05, 0.1, and 0.2 g/100 g feed) of a commercial product containing 92.6 g of ADL per 100 g. The authors observed that lignocellulose at 0.05 and 0.1% improved jejunum mucosal development by increasing villus height and VH/CD ratio, and such a result is in agreement with our findings, since the ADL in the AG group was 0.14% higher than the control. The villus size (height and width) is fundamental for the absorptive activity of the intestine [44]. As a result, a higher VH and CD ratio can be recognized as marker for an increase in the digestion and absorption of nutrients [43]. Moreover, dietary AGs in addition to modifying villus size enhanced the brush border membrane enzymatic activities, contributing to normalizing the nutrient digestion.

The levels of brush border membrane enzymes could have also contributed to normalizing the nutrient digestion despite the lower intestinal length and feed intake of hens fed AGs. The digestion and absorption of almost all the nutrients in the diet take place in the small intestine [45]: around 95% of the fats are digested in the duodenum [46]; fats, starch, and protein are digested in the jejunum and ileum [47], but the ileum is mainly involved in water and mineral absorption [45]. Disaccharidase and peptidase are extrinsic enzymes, which mainly participate in the digestion and absorption of nutrients. An increase in the activities of disaccharidase and peptidase is connected with enhanced nutrient digestion and absorption capacity of the intestinal epithelium [48]. In the duodenum, all the BBM enzyme activities in the AG-fed hens were higher than the control: +109% for maltase, +120 for SI, +108 for L-ANP, and +89.3% for IAP. In the jejunum, only SI (+79.68%) and IAP (+105.8%) were higher in the AG group, while in the ileum, SI showed a lower activity (−72.04%) and IAP a higher activity (+53.6%) in the AG group. Such an increase of the BBM enzyme activity in the hens fed the AGs diet is coherent with the histological observations in the duodenum, where an increase in villus height and villus:crypt ratio was observed. The increased enzyme activity could be related both to a higher enterocyte number and to the quantity and composition of the digesta since malt, SI, and L-ANP are substrate-inducible enzymes. Other features together with the BBM enzyme activity could have affected the absorption of nutrients in hens fed the AGs. Among these, PepT-1 and Na+/K+ ATPase expression and activity, and Na+ availability to co-transport, may have played a role in the substantially equal ileal nutrient digestibility between the experimental diets. Also in Tibaldi et al. [26] the differences related to the BBM enzyme activities were not consistent with the changes in nutrient digestibility in European sea bass.

The different activities of intestinal enzymes are affected, among the other factors, by the diet [49,50]. As the main difference in the two groups was the carbohydrate source, this aspect must be focalized. Even if not measured in the present trial, it is well known that corn is richer in starch, while wheats have higher non-starch polysaccharides (NSPs) contents, which act as anti-nutritional factors in poultry [51]. Accordingly, the AG diet contained 6.98% of hemicelluloses (NDF–ADF) [19] and the control diet only 4.32%. According to De Keyser et al. [52], corn contains 8.74% DM of NSP, and wheat 9.93%. Gebruers et al. [53] stated that emmer contained about half the level of mixed-linkage beta-glucan (0.25–0.45% of DM) present in winter, spring, and spelt wheats (0.50–0.95% of DM). Being indigestible by poultry enzymes, the dietary NSPs undergo a microbial digestion that increases along the gastrointestinal tract, including the upper small intestine [54–57]. The consequent fermentation of NSPs leads to the production of volatile fatty acids and is able to decrease the load of carbohydrate necessary for the development of intestinal enzyme function [58]. However, a decrease of intestinal enzyme function due to the presence of soluble NSPs can be compensated by cellular hyperplasia and hypertrophy [59]. Some studies on commercial NSPs [60] reported higher activities of maltase, sucrase, and alkaline phosphatase in the jejuna of broiler chickens fed diets including xanthan gum. In the ileum, maltase activity was also found marginally increased, thus suggesting that NSPs, other fiber sources, or their metabolites should be involved in the maturation of gut cells, which is crucial for various functions along the villus:crypt axis [58].

The cereal type affected jejunal mucosal disaccharidases. In fact, the increase in intestinal enzymes is mechanically improved when the chyme passes through the digestive tract [61]. As reported by Shakouri et al. [62], the bulk of digesta in the gut of birds fed maize diets, resulting from a higher feed intake than birds fed on barley and wheat, may increase maltase activity.

The increased disaccharidase activity facilitates starch digestion, thus increasing the amount of energy absorbable from feeds and giving to the animals a metabolic advantage, despite the lower amount of dietary starch. Our results indicated that the jejunum and ileum are less sensitive to the effect of the diet, probably because the feeds get there partially digested by the previous intestinal tract. In the jejunum and ileum, the maltase and L-ANP showed similar activity in both the experimental groups. In general, the SI activity was higher in the jejuna and lower in the ilea of the AG group. Our results partially agree with Kohl et al. [50], who observed an effect of dietary starch on maltase activity in the mid intestine of chickens.

The aminopeptidases located in the intestinal brush border, enzymes known to carry out intestinal membrane digestion [27], were significantly affected by AGs inclusion in the hen's diet, while IAP has been traditionally considered a marker of the enterocyte maturation and is involved in the dephosphorylation of the microbial LPS, thus preventing its toxicity [63,64]. These enzymes, in combination with other intestinal membrane-bound enzymes, are important for the absorption of the nutrients in order to keep homeostasis [65]. The increased activity of these intestinal enzymes in the AG-fed hens was consistent with the villi height in the duodenum and jejunum and with the villus:crypt ratio in all the considered intestinal tracts. In previous studies, a modulatory effect of the dietary starch and protein level on the aminopeptidases and disaccharidases activity was not observed in chickens [50]. Therefore, we hypothesize a stimulatory effect of the specific AG composition on the metabolism and development of the enterocytes together with a positive effect on the activity of the IAP and L-ANP.

Furthermore, a high level of IAP activity provides the animals a better defense [63,64].

An additional role of IAP is the dephosphorylation of dietary proteins/peptides, so its high activity in the AG group could be also related to a particular moiety of proteins from the digested AGs.

5. Conclusions

To summarize, based on the histological and physiological analysis of the gut performed in the present study, we suggest that the dietary inclusion of ancient grains is related to positive effects in hens. In particular, the increase of both the digestive enzymes activity and the villus height are indicative of an improved intestinal function without affecting growth and nutrient digestibility, as shown in a previous study [17].

In conclusion, these data characterize the opportunity to include ancient grains in the hen's diet to noticeably affect the structure as well as the overall digestive enzyme activity.

Author Contributions: Conceptualization, F.B. and I.O.; methodology, B.R., M.M. and F.T.; software, F.B.; validation, F.B., M.M., F.T., G.P., A.N.; formal analysis, N.F.A., B.R., M.M.; investigation, N.F.A., N.M., G.P.; data curation, N.F.A., N.M., C.D.M. and F.T.; writing—original draft preparation, F.B., C.D.M., G.P., I.O. and F.T.; writing—review and editing, N.F.A. and N.M. All authors have read and agreed to the published version of the manuscript.

Funding: This research received no external funding.

Institutional Review Board Statement: This research was approved by the Ethical Animal Care and Use Committee of the University of Naples Federico II, Italy (number 2017/0017676). Experiments were per-formed in an organic laying hens farm in Avellino (Italy) for 14 weeks (February–May 2019).

Informed Consent Statement: Not applicable.

Data Availability Statement: Data are available from the authors upon reasonable request.

Conflicts of Interest: The authors declare no conflict of interest.

References

1. Giambanelli, E.; Ferioli, F.; Kocaoglu, B.; Jorjadze, M.; Alexieva, I.; Darbinyan, N.; Antuono, F. A comparative study of bioactive compounds in primitive wheat populations from Italy, Turkey, Georgia, Bulgaria and Armenia. *J. Sci. Food Agric.* **2013**, *93*, 3490–3501. [CrossRef]

2. Shewry, P.R. Do ancient types of wheat have health benefits compared with modern bread wheat? *J. Cereal Sci.* **2018**, *79*, 469–476. [CrossRef]

3. Boukid, F.; Folloni, S.; Sforza, S.; Vittadini, E.; Prandi, B. Current Trends in Ancient Grains-Based Foodstuffs: Insights into Nutritional Aspects and Technological Applications. *Compr. Rev. Food Sci. Food Saf.* **2018**, *17*, 123–136. [CrossRef] [PubMed]

4. Valli, V.; Taccari, A.; Di Nunzio, M.; Danesi, F.; Bordoni, A. Health benefits of ancient grains. Comparison among bread made with ancient, heritage and modern grain flours in human cultured cells. *Food Res. Int.* **2018**, *107*, 206–215. [CrossRef] [PubMed]

5. Zamaratskaia, G.; Gerhardt, K.; Wendin, K. Biochemical characteristics and potential applications of ancient cereals—An underexploited opportunity for sustainable production and consumption. *Trends Food Sci. Technol.* **2021**, *107*, 114–123. [CrossRef]

6. Kulathunga, J.; Reuhs, B.L.; Zwinger, S.; Simsek, S. Comparative Study on Kernel Quality and Chemical Composition of Ancient and Modern Wheat Species: Einkorn, Emmer, Spelt and Hard Red Spring Wheat. *Foods* **2021**, *10*, 761. [CrossRef]

7. Biel, W.; Jaroszewska, A.; Stankowski, S.; Sobolewska, M.; Kępińska-Pacelik, J. Comparison of yield, chemical composition and farinograph properties of common and ancient wheat grains. *Eur. Food Res. Technol.* **2021**, *247*, 1525–1538. [CrossRef]

8. Landberg, R.; Kamal-Eldin, A.; Salmenkallio-Marttila, M.; Rouau, X.; Åman, P. Localization of alkylresorcinols in wheat, rye and barley kernels. *J. Cereal Sci.* **2008**, *48*, 401–406. [CrossRef]

9.   Nirmala Prasadi, V.P.; Joye, I.J. Dietary Fibre from Whole Grains and Their Benefits on Metabolic Health. *Nutrients* **2020**, *12*, 3045.
10.   Thorup, A.C.; Gregersen, S.; Jeppesen, P.B. Ancient Wheat Diet Delays Diabetes Development in a Type 2 Diabetes Animal Model. *Rev. Diabet. Stud.* **2014**, *11*, 245–257. [CrossRef]
11.   Zhang, X.-F.; Wang, X.-K.; Tang, Y.-J.; Guan, X.-X.; Guo, Y.; Fan, J.-M.; Cui, L.-L. Association of whole grains intake and the risk of digestive tract cancer: A systematic review and meta-analysis. *Nutr. J.* **2020**, *19*, 1–14. [CrossRef]
12.   Colombo, F.; Di Lorenzo, C.; Biella, S.; Bani, C.; Restani, P. Ancient and Modern Cereals as Ingredients of the Gluten-Free Diet: Are They Safe Enough for Celiac Consumers? *Foods* **2021**, *10*, 906. [CrossRef]
13.   Hemdane, S.; Jacobs, P.J.; Dornez, E.; Verspreet, J.; Delcour, J.A.; Courtin, C.M. Wheat (Triticum aestivum L.) bran in bread making: A critical review. *Comp. Rev. Food Sci. Food* **2016**, *15*, 28–42. [CrossRef]
14.   Xynias, I.N.; Mylonas, I.; Korpetis, E.G.; Ninou, E.; Tsaballa, A.; Avdikos, I.D.; Mavromatis, A.G. Durum Wheat Breeding in the Mediterranean Region: Current Status and Future Prospects. *Agronomy* **2020**, *10*, 432. [CrossRef]
15.   Arzani, A.; Ashraf, M. Cultivated Ancient Wheats (Triticum spp.): A Potential Source of Health-Beneficial Food Products. *Compr. Rev. Food Sci. Food Saf.* **2017**, *16*, 477–488. [CrossRef]
16.   Guzmán, C.; Alvarez, J.B. Chapter 2—Ancient Wheats Role in Sustainable Wheat Cultivation. In *Trends in Wheat and Bread Making*; Galanakis, C.M., Ed.; Academic Press: Cambridge, MA, USA, 2021; pp. 29–66. ISBN 9780128210482. [CrossRef]
17.   Lombardi, P.; Addeo, N.F.; Panettieri, V.; Musco, N.; Piccolo, G.; Vassalotti, G.; Nizza, A.; Moniello, G.; Bovera, F. Blood profile and productive performance after partial substitution of maize grain with ancient wheat lines by-products in organic laying hens' diet. *Res. Vet. Sci.* **2020**, *133*, 232–238. [CrossRef]
18.   AOAC. *Official Methods of Analysis*, 18th ed.; Association of Official Analytical Chemists: Arlington, VA, USA, 2005.
19.   Van Soest, P.J.; Robertson, J.B.; Lewis, B.A. Methods for dietary fiber, neutral detergent fiber, and nonstarch polysaccharides in relation to animal nutrition. *J. Dairy Sci.* **2020**, *74*, 3583–3597. [CrossRef]
20.   NRC. *Nutrient Requirements of Poultry*, 9 rev. ed.; The National Academies Press: Cambridge, MA, USA, 1994.
21.   Akar, T.; Cengiz2*, M.F.; Tekin1, M. A comparative study of protein and free amino acid contents in some important ancient wheat lines. *Qual. Assur. Saf. Crop. Foods* **2019**, *11*, 191–200. [CrossRef]
22.   Zarantoniello, M.; Bruni, L.; Randazzo, B.; Vargas, A.; Gioacchini, G.; Truzzi, C.; Annibaldi, A.; Riolo, P.; Parigi, G.; Cardinaletti, G.; et al. Partial Dietary Inclusion of Hermetia illucens (Black Soldier Fly) Full-Fat Prepupae in Zebrafish Feed: Biometric, Histological, Biochemical, and Molecular Implications. *Zebrafish* **2018**, *15*, 519–532. [CrossRef]
23.   Moniello, G.; Ariano, A.; Panettieri, V.; Tulli, F.; Olivotto, I.; Messina, M.; Randazzo, B.; Severino, L.; Piccolo, G.; Musco, N.; et al. Intestinal Morphometry, Enzymatic and Microbial Activity in Laying Hens Fed Different Levels of a Hermetia illucens Larvae Meal and Toxic Elements Content of the Insect Meal and Diets. *Animals* **2019**, *9*, 86. [CrossRef]
24.   Zarantoniello, M.; Randazzo, B.; Truzzi, C.; Giorgini, E.; Marcellucci, C.; Vargas-Abùndez, J.A.; Zimbelli, A.; Annibaldi, A.; Parigi, G.; Tulli, F.; et al. A six-months study on Black Soldier Fly (Hermetia illucens) based diets in zebrafish. *Sci. Rep.* **2019**, *9*, 8598. [CrossRef] [PubMed]
25.   Shirazy-Beechey, S.P.; Hirayama, B.A.; Wang, Y.; Scott, D.; Smith, M.W.; Wright, E.M. Ontogenic development of lamb intestinal sodium glucose co-transporter is regulated by diet. *J. Physiol.* **1991**, *437*, 699–708. [CrossRef] [PubMed]
26.   Tibaldi, E.; Hakim, Y.; Uni, Z.; Tulli, F.; de Francesco, M.; Luzzana, U.; Harpaz, S. Effects of the partial substitution of dietary fish meal by differently processed soybean meals on growth performance, nutrient digestibility and activity of intestinal brush border enzymes in the European sea bass (Dicentrarchus labrax). *Aquaculture* **2006**, *261*, 182–193. [CrossRef]
27.   Vizcaíno, A.J.; López, G.; Sáez, M.I.; Jiménez, J.A.; Barros, A.; Hidalgo, L.; Camacho-Rodríguez, J.; Martínez, T.F.; Cerón-García, M.C.; Alarcón, F.J. Effects of the microalga Scenedesmus almeriensis as fishmeal alternative in diets for gilthead sea bream, Sparus aurata, juveniles. *Aquaculture* **2014**, *431*, 34–43. [CrossRef]
28.   Bradford, M.M. A rapid and sensitive method for the quantification of microgram quantities of protein utilizing the principle of protein-dye binding. *Ann. Biochem.* **1976**, *72*, 248–254. [CrossRef]
29.   SAS. *Statistical Analyses System. SAS/STAT Software, Version 9*; SAS Institute Inc.: Cary, NC, USA, 2000.
30.   El-Katcha, M.I.; El-Kholy, M.E.; Soltan, M.A.; EL-Gayar, A.H. Effect of Dietary Omega-3 to Omega-6 Ratio on Growth Performance, Immune Response, Carcass Traits and Meat Fatty Acids Profile of Broiler Chickens. *Poultry Sci.* **2014**, *2*, 71–94.
31.   Awad, W.A.; Hess, M.; Twarużek, M.; Grajewski, J.; Kosicki, R.; Böhm, J.; Zentek, J. The impact of the Fusarium mycotoxin deoxynivalenol on the health and performance of broiler chickens. *Int. J. Mol. Sci.* **2011**, *12*, 7996–8012. [CrossRef]
32.   Hetland, H.; Choct, M.; Svihus, B. Role of insoluble non-starch polysaccharides in poultry nutrition. *World's Poult. Sci. J.* **2004**, *60*, 415–422. [CrossRef]
33.   Smits, C.H.M.; Annison, G. Non-starch plant polysaccharides in broiler nutrition—Towards a physiologically valid approach to their determination. *World's Poult. Sci. J.* **2019**, *52*, 203–221. [CrossRef]
34.   Cavazos, A.; Gonzalez de Mejia, E. Identification of Bioactive Peptides from Cereal Storage Proteins and Their Potential Role in Prevention of Chronic Diseases. *Compr. Rev. Food Sci. Food Saf.* **2013**, *12*, 364–380. [CrossRef]
35.   Shewry, P.R.; Halford, N.G. Cereal seed storage proteins: Structures, properties and role in grain utilization. *J. Exp. Bot.* **2002**, *53*, 947–958. [CrossRef]
36.   Imondi, A.R.; Bird, F.H. The sites of nitrogen absorption from the alimentary tract of the chicken. *Poult. Sci.* **1965**, *44*, 916–920. [CrossRef]

37. Yamauchi, K.; Yamamoto, K.; Isshiki, Y. Morphological alterations of the intestinal villi and absorptive epithelial cells in each intestinal part in fasted chickens. *Jpn. Poult. Sci.* **1995**, *32*, 241–251. [CrossRef]
38. Yamauchi, K.; Kamisoyama, H.; Isshiki, Y. Effect of fastingand refeeding on structures of the intestinal villi and epithelial cell in While Leghorn hens. *Br. Poult. Sci.* **1996**, *37*, 909–921. [CrossRef]
39. Svihus, B.; Juvik, E.; Hetland, H.; Krogdahl, A. Causes for Improvement in Nutritive Value of Broiler Chicken Diets with Whole Wheat Instead of Ground Wheat. *Br. Poult. Sci.* **2004**, *45*, 55–60. [CrossRef]
40. Zimonja, O.; Svihus, B. Effects of processing of wheat or oats starch on physical pellet quality and nutritional value for broilers. *Anim. Feed Sci. Technol.* **2009**, *149*, 287–297. [CrossRef]
41. Yamauchi, K. Review of a histological intestinal approach to assessing the intestinal function in chickens and pigs. *Anim. Sci. J.* **2007**, *78*, 356–370. [CrossRef]
42. Samanya, M.; Yamauchi, K. Morphological change of the intestinal villi in chicken fed dietary charcoal powder, including wood vinegar compound. *J. Poult. Sci.* **2001**, *38*, 289–301. [CrossRef]
43. Sozcu, A.; Ipek, A. The effects of lignocellulose supplementation on laying performance, egg quality parameters, aerobic bacterial load of eggshell, serum biochemical parameters, and jejunal histomorphological traits of laying hens. *Poult. Sci.* **2020**, *99*, 3179–3187. [CrossRef]
44. Iji, P.A.; Saki, A.A.; Tivey, D.R. Intestinal development and body growth of broiler chicks on diets supplemented with non-starch polysaccharides. *Anim. Feed Sci. Technol.* **2001**, *89*, 175–188. [CrossRef]
45. Svihus, B. The gizzard: Function, influence of diet structure and effects on nutrient availability. *World's Poult. Sci. J.* **2011**, *67*, 207–224. [CrossRef]
46. Noy, Y.; Sklan, D. Digestion and absorption in the young chick. *Poult. Sci.* **1995**, *74*, 366–373. [CrossRef] [PubMed]
47. Riesenfeld, G.; Sklan, D.; Bar, A.; Eisner, U.; Hurwitz, S. Glucose absorption and starch digestion in the intestine of the chicken. *J. Nutr.* **1980**, *110*, 117–121. [CrossRef] [PubMed]
48. Moreno, M.J.; Pellicer, S.; Fernandezotero, M.P. Lindane treatment alters both intestinal-mucosa composition and brush-border enzymatic-activity in chickens. *Pestic. Biochem. Phys.* **1995**, *52*, 212–221. [CrossRef]
49. Kenny, A.J.; Turner, A.J. What are ecto-enzymes? In *Mammalian Ecto-Enzymes*; Elsevier: New York, NY, USA; Oxford, MS, USA, 1987; pp. 2–13.
50. Kohl, K.D.; Ciminari, M.E.; Chediack, J.G.; Leafloor, J.O.; Karasov, W.H.; McWilliams, S.R.; Caviedes-Vidal, E. Modulation of digestive enzyme activities in the avian digestive tract in relation to diet composition and quality. *J. Comp. Physiol. B* **2017**, *197*, 339–351. [CrossRef] [PubMed]
51. Choct, M.; Hughes, R.J.; Trimble, R.P.; Angkanaporn, K.; Annison, G. Non-starch polysaccharide-degrading enzymes increase the performance of broiler chickens fed wheat of low apparent metabolizable energy. *J. Nutr.* **1995**, *125*, 485–492. [PubMed]
52. De Keyser, K.; Dierick, N.; Kuterna, L.; Maigret, O.; Kaczmarek, S.; Rutkowski, A.; Vanderbeke, E. Non-starch Polysaccharide Degrading Enzymes in Corn and Wheat-Based Broiler Diets: Dual Activity for Major Substrates. *J. Agric. Sci. Tech.* **1987**, *8*, 76–88.
53. Gebruers, K.; Dornez, E.; Boros, D.; Fraś, A.; Dynkowska, W.; Bedo, Z.; Rakszegi, M.; Delcour, J.A.; Courtin, C.M. Variation in the content of dietary fiber and components thereof in wheats in the HEALTHGRAIN Diversity Screen. *J. Agric. Food Chem.* **2008**, *56*, 9740–9749. [CrossRef]
54. Ricke, S.C.; Van der Aar, P.J.; Fahey, G.C.J.; Berger, L.L. Influence of dietary fibers on performance and fermentation characteristics of gut contents from growing chicks. *Poult. Sci.* **1982**, *61*, 1335–1343. [CrossRef]
55. Choct, M.; Hughes, R.J.; Wang, J.; Bedford, M.R.; Morgan, A.J.; Annison, G. Increased small intestinal fermentation is partly responsible for the anti-nutritive activity of non-starch polysaccharides in chickens. *Br. Poult. Sci.* **1996**, *37*, 609–621. [CrossRef]
56. Jørgensen, H.; Zhao, X.Q.; Knudsen, K.E.B.; Eggum, B.O. The influence of dietary fibre source and level on the development of the gastrointestinal tract, digestibility and energy metabolism in broiler chickens. *Br. J. Nutr.* **1996**, *75*, 379–395. [CrossRef]
57. Smits, C.H.; Veldman, A.; Verstegen, M.W.A.; Beynen, A.C. Dietary carboxymethylcellulose with high instead of low viscosity reduces macronutrient digestion in broiler chickens. *J. Nutr.* **1997**, *127*, 483–487. [CrossRef]
58. Iji, P.A. The impact of cereal non-starch polysaccharides on intestinal development and function in broiler chickens. *World's Poult. Sci. J.* **1999**, *55*, 375–387. [CrossRef]
59. Ikegami, S.; Tsuchihashi, F.; Harada, H.; Nishide, E.; Innami, S. Effect of viscous indigestible polysaccharides on pancreatic-bilary secretion and digestive organs in rats. *J. Nutr.* **1990**, *120*, 353–360. [CrossRef]
60. Iji, P.A.; Tivey, D.R. Natural and synthetic oligosaccharides in broiler chicken diets. *World's Poult. Sci. J.* **1998**, *54*, 129–143. [CrossRef]
61. Duke, G.E. Recent studies on regulation of gastric motility in turkeys. *Poult. Sci.* **1992**, *71*, 1–8. [CrossRef]
62. Shakouri, M.D.; Iji, P.A.; Mikkelsen, L.L.; Cowieson, A.J. Intestinal function and gut microflora of broiler chickens as influenced by cereal grains and microbial enzyme supplementation. *J. Anim. Physiol. Anim. Nutr.* **2009**, *93*, 647–658. [CrossRef]
63. Lallès, J.P. Intestinal alkaline phosphatase: Multiple biological roles in maintenance of intestinal homeostasis and modulation by diet. *Nutr. Rev.* **2010**, *68*, 323–332. [CrossRef]
64. Lallès, J.P. Intestinal alkaline phosphatase: Novel functions and protective effects. *Nutr. Rev.* **2014**, *72*, 82–94. [CrossRef]
65. National Research Council. *Nutrient Requirements of Fish and Shrimp*; The National Academies Press: Washington, DC, USA, 2011.

 *sustainability*

Article

Optical Characterization of *Alternaria* spp. Contaminated Wheat Grain and Its Influence in Early Broilers Nutrition on Oxidative Stress

Nikola Puvača [1,*] , Snežana Tanasković [2] , Vojislava Bursić [3], Aleksandra Petrović [3], Jordan Merkuri [4], Tana Shtylla Kika [5], Dušan Marinković [3], Gorica Vuković [6] and Magdalena Cara [4]

[1] Department of Engineering Management in Biotechnology, Faculty of Economics and Engineering Management in Novi Sad, University Business Academy in Novi Sad, Cvećarska 2, 21000 Novi Sad, Serbia
[2] Faculty of Agronomy in Čačak, University of Kragujevac, Cara Dušana 34, 32102 Čačak, Serbia; stanasko@kg.ac.rs
[3] Department for Phytomedicine and Environmental Protection, Faculty of Agriculture, University of Novi Sad, Trg Dositeja Obradovića 8, 21000 Novi Sad, Serbia; bursicv@polj.uns.ac.rs (V.B.); aleksandra.petrovic@polj.uns.ac.rs (A.P.); dusan.marinkovic@polj.uns.ac.rs (D.M.)
[4] Department of Plant Protection, Faculty of Agriculture and Environment, Agricultural University of Tirana, Koder Kamez, 1029 Tirana, Albania; jordanmerkuri@gmail.com (J.M.); mcara@ubt.edu.al (M.C.)
[5] Faculty of Veterinary Medicine, Agricultural University of Albania, Kodër Kamëz, SH1, 1000 Tirana, Albania; tana_kika@hotmail.com
[6] Faculty of Agriculture, University of Belgrade, Nemanjina 6, 11080 Belgrade-Zemun, Serbia; goricavukovic@yahoo.com
* Correspondence: nikola.puvaca@fimek.edu.rs; Tel.: +381-65-219-1284

 **check for updates**

Citation: Puvača, N.; Tanasković, S.; Bursić, V.; Petrović, A.; Merkuri, J.; Shtylla Kika, T.; Marinković, D.; Vuković, G.; Cara, M. Optical Characterization of *Alternaria* spp. Contaminated Wheat Grain and Its Influence in Early Broilers Nutrition on Oxidative Stress. *Sustainability* **2021**, *13*, 4005. https://doi.org/10.3390/su13074005

Academic Editor: George K. Symeon

Received: 8 March 2021
Accepted: 1 April 2021
Published: 3 April 2021

Publisher's Note: MDPI stays neutral with regard to jurisdictional claims in published maps and institutional affiliations.

Abstract: The aim of this research was the visual characterization and investigating the effects of *Alternaria* spp. contaminated wheat grains in the starter stage of broilers nutrition on productive parameters and oxidative stress. The research was divided into two phases. Bunches of wheat in post-harvest period of year 2020 was collected from a various locality in Serbia and Albania. In the first phase, collected samples were visual characterized by *Alternaria* spp. presence by color measurement methods. Gained results are conferred in the range of the color properties of grain color properties of *Alternaria* toxins. Wheat grain samples were significantly different ($p < 0.05$) in terms of all measured color parameters (L^*, a^*, b^*). Classification of field fungi in analyzed wheat grain samples showed that the significant field fungi were *Rhizopus* spp., followed by *Alternaria* spp., and *Fusarium* spp. In the second phase, biological tests with chickens were carried out during the broiler chickens' dietary starter period in the first 14th days of age. At the beginning of the experiment, a total of 180-day-old Ross 308 strain broilers were equally distributed into three dietary treatments, with four replicates each. Dietary treatments in the experiments were as follows: basal diet without visual contamination of *Alternaria* spp. with 25% wheat (A1), a basal diet with visual contamination of *Alternaria* spp. with 25% wheat from Serbia (A2), basal diet with visual contamination of *Alternaria* spp. with 25% wheat from Albania (A3). The trial with chickens lasted for 14 days. After the first experimental week, wheat infected with *Alternaria* spp. in treatment A2 and A3 expressed adverse effects. The highest body weight of chickens of 140.40 g was recorded in broilers on control treatment A1 with statistically significant differences ($p < 0.05$) compared to treatments A2 (137.32 g) and A3 (135.35 g). At the end of the second week of test period, a statistically significant ($p < 0.05$) difference in body weight of broiler chickens could be noticed. The highest body weight of 352.68 g was recorded in control treatment A1, with statistically significant differences compared to other *Alternaria* spp. treatments. The lowest body weight of chickens was recorded in treatment A3 (335.93 g). Results of feed consumption and feed conversion ratio showed some numerical differences between treatments but without any statistically significant differences ($p > 0.05$). *Alternaria* spp. contaminated diet increased glutathione (GSH), glutathione reductase (GR), alanine aminotransferase (ALT), and aspartate aminotransferase (AST) and decreased peroxidase (POD) and superoxide dismutase (SOD) serum levels, respectively. Built on the achieved results, it can be concluded that the wheat contaminated with *Alternaria* spp. in broilers nutrition negatively affected growth, decreased oxidative protection and interrupted chicken welfare in the first period of life.

Keywords: *Alternaria* spp.; mycotoxins; fungi; poultry production; welfare; oxidative stress

1. Introduction

Corn and wheat represent the primary energy source in the food animal's daily diet, while wheat has been considered the third most-produced feedstuff globally [1]. In the last ten years, studies and researchers have been struggling with the fungi of the genus *Alternaria*, which has grown to be the leading cause of wheat grains contamination [2]. The essential characteristics of *Alternaria* genera is the production of melanin and the host-specific plant–fungi/toxin interaction [3–5]. In addition, direct melanin emerges an indirect role in virulence, as well [6]. Melanin poses the ability to function as the shield in plant fungi protections versus ecological stress or unfavorable conditions, which gives fungus permanency and endurance. Furthermore, melanin promptly responds with free oxygen radicals, versus the pathogen's infiltration in the plant-host cells [7,8]. The blackening of the wheat grain lobes prior to cropping is typical indicator of contamination with *Alternaria* spp. [9]. At hand there is several forms of discoloration that can alter ordinary wheat (*Triticum aestivum* L.). In nearly all areas where wheat is cultivated, the black point is usually correlated by *Alternaria alternata* as a common discoloration of seed [10]. The staining usually appears in the external pericarp and internal grain tissue and could broaden beside its adaxial side. Such kinds of grain color changes differ drastically in frequency and seriousness depending on grain during the maturation. Biotic and abiotic stresses can cause wheat grain color changes, often caused by high humidity and high temperatures [11]. Those kinds of conditions are very favorable for fungi and mycotoxins occurrence in general [12,13]. Recently it has been confirmed that high humidity levels might stimulate the sporadic expansion of black point on wheat grain under controlled conditions [14]. *A. alternata* was the primary cause related to black point occurrence on wheat grain [15]. Likewise, pathogenicity and decrease of quality of wheat grains are influenced by a number of *Alternaria* spp. the producers of toxic secondary metabolites known as *Alternaria* mycotoxins [16,17]. *Alternaria* mycotoxins as alternariol (AOH) [18], tenuazonic acid (TzA) [19], alternariol monomethyl ether (AME) [20], altenuene (ALT) [2], altertoxin I (ATX-I) [21], alterotoxin II (ATX-II) [18], and stemphyltoxin III (STTX-III) [22] could be toxic for animal health [23].

Some of the previous mentioned toxins could cause a serious health damages in animals when ingested, between them, for instance, fetotoxicity and somatic or functional deficiencies in the fetus when the mother is exposed to toxins [23]. *A. alternata*, as a separate mycotoxin, is mutagenic and clastogenic in various in vitro systems [24]. Moreover, it has been recommended that *Alternaria* toxins in grains be accountable for gullet pipe cancer [25]. Consequently, because of toxic effects, *Alternaria* toxins are of concern for public and animal health [26]. The European Commission (EC), and European Food Safety Authority (EFSA) were therefore engaged to give a technical view on the hazards for community and animal wellbeing associated with the occurrence of *Alternaria* mycotoxins in the commodities for human and animal daily nutrition. Subsequently, *A. alternata* have been chemically characterized, and incidence in feed was recorded [27]. Nevertheless, more than a few other *Alternaria* toxins have been classified as well, respectively [28].

Assessment of *Alternaria* toxins consumption by food animals through daily feeding have been restricted to broilers since poultry have been single one animal race where certain information about mycotoxin toxicity is appropriate for hazard evaluation [29–31]. Given that the incidence of feed data was lacking for the majority of the *Alternaria* toxins, the exposure assessments have been restricted to AOH toxin. Estimated lower bound and upper bound introductions to alternariol (AOH) were around 0.003 mg/day and 0.006 mg/day, for chickens and layers, respectively.

Broilers in production conditions are subjected to a variety of stressors [32]. The additional reactive oxygen species (ROS) and reactive nitrogen species (RNS) production and oxidative stress are the essential harmful outcomes [33]. In the evolutionary process, antioxidant defense mechanism were built in birds to be able to stay alive in an oxygenated atmosphere [34]. They consist of a dense system of inside integrated antioxidant enzymes, for instance, glutathione (GSH), coenzyme Q (CoQ), and outwardly provided by vitamins, carotenoids, and antioxidants [34]. Furthermore, all antioxidants in the body work together to sustain the best oxidoreduction equilibrium [35]. This equilibrium is a crucial component in supplying the required preconditions for cells indicating, stress adjustment, and homeostasis upkeep [36]. While ROS and RNS are critical signaling molecules, their presence have been rigorously controlled by the antioxidant defense system linked with various transcript components and vitagenes [37]. Physiology shows that change from optimum inner and outer circumstances causes stress [38].

Additionally, a complicated flow of controlling systems is implicated in the stress reaction, causing the metabolic alterations triggering weakened live performance in broilers [39]. When the ROS and RNS construction outstrips the antioxidant defense mechanism ability to neutralize them, oxidative stress arises [38]. That includes polyunsaturated fatty acids (PAFAs), proteins, and DNA [40], take the lead to damaging outcomes in wellbeing, progress, development, and overall animal welfare [41].

Contemplating lucking research results and significant information's on *Alternaria* mycotoxins and that the biochemical composition of more than a few is identified, this research's precise aim was to visually characterize and investigate the effects of *Alternaria* spp. contaminated wheat grains in broiler chicken nutrition in the starter stage on productive parameters, oxidative stress, and overall welfare of this species of food animals. Obtained results from this research can serve in the future as the reference material for creating the new up-to-date guidelines on *Alternaria* toxins in foodstuffs and feedstuffs.

2. Materials and Methods

2.1. Wheat Samples

Wheat grain samples (*Triticum aestivum*) were collected in post-harvest time in the season of 2020 from the region of Serbia (Vojvodina) and Albania (Durrës). Obtained samples were collected with the appropriate equipment, such as a probe for stationary grain and a diverter-type mechanical sampler, using a sampling pattern and procedures designed to collect samples from all areas of the lot. The appropriate size of wheat grain between 1.5 and 2.5 kg sample was taken from a truck with adequate identified and labeled bags. Collected samples were handled in such a way as to maintain representativeness. Samples were stored in a cool and dry place in triple lined paper breathable bags to avoid mold growth and increase of sample moisture level over 14%.

2.2. Proximate Analyses of Compound Feed for Broilers Chickens

Compound feed for broiler chickens in each experimental treatment were analyzed for moisture, crude ash, crude protein, crude fat, and crude fiber. All analyses were performed in triplicate. The moisture content was determined according to AOAC (Association of Official Analytical Chemists) [42] Method 934.01. Crude protein content was determined by Kjeldahl method according to the AOAC Method 978.04, crude ash, according to AOAC Method 942.05, crude fat, according to AOAC Method 920.39 and crude fiber according to AOAC Method 978.10 (AOAC, 1998). Concentration of total phosphorus (P) and calcium (Ca) as well as metabolizable energy (ME), was calculated within licenced feed formulation software.

2.3. Visual Scale Establishing and Color Measurement

Instrumental methods were used to measure the color of wheat grains. The wheat grain samples color was measured with Minolta Chroma Meter CR-400, and the attachment CR-A50, respectively. The color space defined by the International Commission on Illumination (CIELAB) L^* (lightness), a^* (red-green), and b^* (yellow-blue), and dominant wavelength (DWL) was determined using a $D65$ light source and the observer angle at

2°. The tristimulus values of L^*, a^* and b^* readings were calibrated against a standard white plate (Y = 84.8; x = 0.3199; y = 0.3377). Each wheat ear sample was divided into four subgroups, and the color of one hundred ears from each subgroup (400 ears from one sample) was measured on 5 locations. Samples of wheat grains was divided into 4 subgroups, and 5 repetitions measure were applied in each subgroup (20 repetitions per sample in total) [43].

2.4. Wheat Grain Infection Confirmation

Precisely one hundred wheat grains were counted randomly and used in all the treatments. Therefore, wheat grain samples must be cleaned with 0.4% NaOCl, and washed with clean water for 2 min. After cleaning process samples of wheat was placed on Petri dishes in 4 replicates (25 grains/Petri) containing potato dextrose culture medium. Grains were incubated at 25 °C, in total of 7 days, after which concentration of contamination was evaluated. For validation of fungi species from each Petri dish 5 randomly wheat grain was taken out by microscopic inspection after finishing the incubation period [44].

2.5. In Vivo Experimental Part with Broiler Chickens

The test with the broilers was conducted in the wake of the EU legislation and tenet of the 3Rs contained by Directive 2010/63/EU, as well with the approval of Ethic Commission for the Protection and Welfare of Experimental Animals EK-I-2020-01. At the beginning of the investigation, a total of 180-day-old Ross 308 strain broilers were equally distributed into three dietary treatments, with four replicates each. Dietary treatments in the experiments were as follows: basal diet without visual contamination of *Alternaria* spp. with 25% wheat (A1), a basal diet with visual contamination of *Alternaria* spp. with 25% wheat from Serbia (A2), and basal diet with visual contamination of *Alternaria* spp. with 25% wheat from Albania (A3). Wheat contamination of *Alternaria* spp. from control treatment A1 was prepared as $1/2$ mixture of wheat samples from both Serbia and Albania. During the test period broilers was given feed and water by will e.g., ad libitum, with regularly monitored and maintained environmental conditions provided by broilers producer. Broilers were kept on the ground bedding system with the pelleted wheat straw. To control the productive results of broilers, body weight, feed consumption, and feed utilization were monitored.

2.6. Blood Samples Collection and Hemolysate Preparation

The broiler chickens' blood was collected by the trained veterinarian from the broilers' heart by puncture into heparinized sterile tubes. Blood samples were immediately delivered to the laboratory and centrifuged for 10 min at 1507 g and 4 °C. Plasma was removed, following the erythrocytes rinsing in saline three times. The obtained red blood cell pellet was held in same amount of two filtered water and vortexed afterward. Following incubation for 60 min at 25 °C, the hemolysate was centrifuged during fifteen minutes at 1507 g, after which obtained buoyant was gathered for additional examination [45].

2.7. Determination of Glutathione and Enzymatic Determination

To determine glutathione (GSH) concentration, proteins from hemolysates were divided by increasing half amount of 10% sulfosalicylic acid and centrifuged at 3075 g, for five minutes, at 4 °C. The buoyant was deposited at 4 °C, and GSH was determined the next day. The GSH concentration in the blood hemolysate was determined from the quantity of sulfhydryl residues [45].

Superoxide dismutase (SOD) activity was determined by the spectrophotometric method based on the inhibition of adrenaline reduction to adrenochrome at pH 10.2 [45]. The activity of glutathione reductase (GR) was determined from the rate of nicotinamide adenine dinucleotide phosphate (NADPH) oxidation measured at the absorbance at 340 nm [46]. The concentration of lipid peroxides (LPx) was determined by the thiobarbituric acid (TBA) assessment [47]. The oxidation of cellular membrane lipids was measured through the reaction of lipid peroxides with TBA [47]. The determination of peroxidase (POD) activity was based on the catalytic oxidation of guayacol by H_2O_2 as an electron acceptor [45]. The reaction of xanthine oxidation of uric acid was used for the determination of xanthine oxidase (XOD) activity. Spectrophotometric measurement was performed in 0.1 mmol/dm^3 phosphate buffer at pH 7.5, at the absorbance at 295 nm [45].

2.8. Serum Biochemical Analyses

The serum activities of aspartate aminotransferase (AST) and alanine aminotransferase (ALT) were determined in serum samples. Analysis of the serum samples was measured by an automatic biochemistry analyzer (Beckman Synchron CX4 PRO, Fullerton, CA, USA) [48].

2.9. Statistical Analyses of Data

The data acquired in the conducted examination were evaluated by one-way analysis of variance (ANOVA) using the software package Statistica 13. Once the analysis of variance exhibited statistical significance, Duncan's MRT was employed. A significant difference was registered at $p < 0.05$.

3. Results and Discussion

Results of proximate analysis of compound feed used in daily nutrition of broiler chickens during the experiment are presented in Table 1.

Table 1. Proximate composition and diet ingredients of compound feed, %.

Nutrients	Treatments		
	A1—Control	A2—Serbia	A3—Albania
Dry matter	89.6	89.5	89.6
Moisture	10.4	10.5	10.4
Crude protein	22.0	22.1	22.1
Crude fat	5.1	5.0	5.2
Crude fiber	3.5	3.4	3.5
Crude ash	6.3	6.4	6.2
Ca	1.0	1.1	0.9
P	0.8	0.8	0.8
Metabolizable Energy, MJ/kg *	12.5	12.5	12.5
Diet ingredients			
Corn	35.4	35.6	35.5
Wheat	25.0	25.0	25.0
Soybean meal	19.5	19.5	19.5
Sunflower meal	2.0	2.0	2.0
Soy protein isolate	8.8	8.6	8.5
Corn gluten	2.0	2.0	2.0
Yeast	1.5	1.5	1.5
Limestone	1.8	1.8	1.8
Premix	4.0	4.0	4.0

* Values were calculated.

Gained results are conferred in the range of the color properties of grain color properties of *Alternaria*. Gained results are conferred in the range of the color properties of *Alternaria* spp. contaminated wheat grains [49]. Wheat grain samples were significantly different ($p < 0.05$) in terms of all measured color parameters (L^*, a^*, b^*). Control wheat grain (A1) samples were significantly different in terms of lightness and dominant wavelength, compared to wheat grain samples (A2) and (A3), which have shown significant difference ($p < 0.05$) compared to A1, but without any statistically significant difference ($p > 0.05$) between themselves, nevertheless numerical differences (Figure 1), respectively.

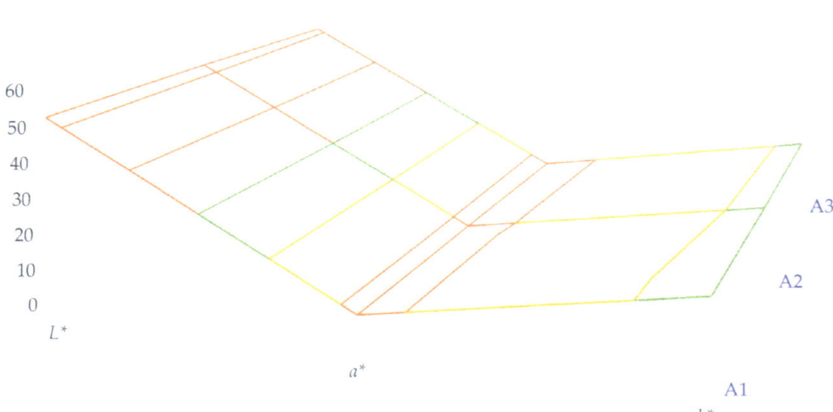

Figure 1. Color parameters of different samples of wheat grains. [A1]—wheat grain without visual contamination of *Alternaria* spp.; [A2]—wheat grain with visual contamination of *Alternaria* spp. (Serbia); [A3]—wheat grain with visual contamination of *Alternaria* spp. (Albania); [L*]—lightness; [a*]—red/green value; [b*]—blue/yellow value.

The results presented in Figure 1 show that all wheat grain samples belong to the different groups by dominant wavelength values. Contemplating all stated, it can be concluded that infection entered the grain in a higher amount in some wheat samples (A2 and A3). Simultaneously, there were samples without visible infection and color changes on the grain (A1). Wheat grain samples without visible dark spots were commonly described by higher lightness and more prominent yellow tones [50].

All wheat grain samples collected from the field and previously instrumentally analyzed were disinfected with 0.4% NaOCl and placed for incubation (Section 2.4) for seven days. Results of fungi genera confirmation were carried out by microscopic examination, and the results have been shown in Table 2.

Table 2. Incidence of some genera of fungi in wheat grain samples, %.

Treatment/Sample	Fungi spp.			Other Fungi spp.
	Alternaria	*Rhizopus*	*Fusarium*	Not Identified
A1—Control	25.2 [b]	72.2 [a]	1.0 [b]	1.6
A2—Serbia	37.9 [a]	48.3 [b]	3.5 [a]	10.3
A3—Albania	39.2 [a]	46.6 [b]	3.9 [a]	10.3
p-value	0.023	0.016	0.003	

Values in the same row marked with the different letters are significantly different at the significance level of $p < 0.05$.

Classification of field fungi in analyzed wheat grain samples showed that the significant field fungi were *Rhizopus* spp., followed by *Alternaria* spp., and *Fusarium* spp. The ratio of contamination of wheat grain samples by *Alternaria* spp. was the highest in A3 samples without significant difference ($p > 0.05$) compared to A2, as previously stated. Differences in percentages between A2 and A3 could be explained by the fact that *Alternaria* spp. produce melanin pigments of dark color, which can cause the differentiation in determination with instrumental measurement, respectively. Due to fungi growth in the field even at low temperatures, they are also responsible for spoilage of commodities during refrigerated transport and storage. Several *Alternaria* species are known producers of toxic secondary metabolites known as *Alternaria* mycotoxins [23]. *A. alternata* produces several mycotoxins. TeA is harmful to several animal species, e.g., mice, chickens, and dogs [23]. Many *Alternaria* metabolites have been reported to occur naturally in cereals [5,12]. Alternariol, alternariol monomethyl ether, and tenuazonic acid were frequently detected in sorghum, wheat, and edible oils [23]. Xu et al. [51] have reported the importance and danger of exposure to *Alternaria* toxins from grain and grain-based products because of its relation to human esophageal cancer in China. In their study, a total of 370 freshly harvested wheat grain samples were analyzed for the four *Alternaria* toxins TeA, TEN, AOH, and AME. Field contaminated samples (95%) of the wheat grains were positive for more than one type of *Alternaria* toxins [51]. Li and Yoshizawa [52] reported the first report of the natural occurrence of *Alternaria* mycotoxins in Chinese wheat. Their wheat grains were significantly infested by *Alternaria* species, mainly *A. alternata*, with a median infection rate of 87.3%. The grains with low quality which is acceptable in some cases was researched in post-harvest period to investigate if the *Alternaria* or *Fusarium* influenced in adverse quality of the grains [53]. The distribution of *Alternaria* and *Fusarium* spp. they were varied significantly in samples of reduced rate compared with acceptable samples. The results of Kosiak et al. [53] revealed a negative interaction between *F. graminearum* and *Alternaria* spp. as well as between *F. graminearum* and another *Fusarium* spp. *Fusarium* and *Alternaria* fungi naturally occurring on the ears and the formation of their mycotoxins in the ripe grains. Müller et al. [9] investigated the fluorescent pseudomonads colonizing wheat ears, which have a high antagonistic potential against phytopathogenic fungi. Unfortunately, the results of their findings indicate that extensive biological management of mycotoxin development by naturally arising pseudomonads with incompatible activity is very doubtful [9].

Based on the gained results in the second phase of the experiment with the live broiler chickens, after the first experimental week, it could be noticed that the addition of wheat infected with *Alternaria* spp. in the amount of 25% in treatment A2 and A3 expressed adverse effects. The highest body weight of chickens of 140.40 g was recorded in broilers on control treatment A1 with statistically significant differences ($p < 0.05$) compared to treatments A2 (137.32 g) and A3 (135.35 g).

At the end of the second week of test period, a statistically significant ($p < 0.05$) difference in body weight of broiler chickens could be noticed. The highest body weight of 352.68 g was recorded in control treatment A1, with statistically significant differences compared to other *Alternaria* spp. treatments. The lowest body weight of chickens was recorded in treatment A3 (335.93 g), while significant differences ($p > 0.05$) between chickens in *Alternaria* spp. treatments were not recorded (Table 3). The low broiler chicken body weight observed in *Alternaria* spp. contaminated diet than control could be due to *Alternaria* spp. toxin tenuazonic acid which was firstly described in 1987 [54].

Table 3. Broiler chickens body weight in the experiment, g.

Age	Treatments in Test			Pooled	
	A1	A2	A3	SE	p
0 day	35.52 [a] ± 2.82	35.38 [a] ± 2.66	34.97 [a] ± 2.77	0.16	0.098
7 day	140.40 [a] ± 9.61	137.32 [b] ± 8.13	135.35 [b] ± 8.19	0.47	0.000
14 day	352.68 [a] ±18.44	341.85 [b] ± 23.30	335.93 [b] ± 22.42	1.29	0.000

Values in the same row marked with the different letters are significantly different at the significance level of $p < 0.05$.

Numerous researches have registered a broad array of serious wellbeing impacts and medical indications after food animals was subjected to the elevated amount of toxins. Nevertheless, not a lot is seen concerning the wellbeing impacts of toxins at small amounts [55]. Kolawole et al. [55] conducted a long-term feeding trial in order to investigate the impact of small amounts of toxin combinations on the production of poultry fed with naturally contaminated complete feed. Total of eighteen tests with poultry production was performed, with closely of 2200 one-day-old Ross-308 birds per each test. As food animals are frequently subjected to low doses of mycotoxin, a cumulative risk evaluation in quantifying and alleviating counter to the economic, welfare, and health influences is necessary for mycotoxins. Hessel-Pras et al. [56] stated that once *Alternaria* mycotoxins passes the intestinal barrier, they can reach the liver to exert yet uncharacterized molecular effects. Hence, the same group of authors used hepatic in vitro systems to examine selected *Alternaria* mycotoxins for their induction of metabolism-dependent cytotoxicity, phosphorylation of the histone H_2AX surrogate marker for DNA double-strand breaks, and relevant marker genes for hepatotoxicity. They have found evidence that 50 μmol/L of AOH, AME, TeA, and TEN might be associated with hepatotoxic effects, necrosis, and the development of diseases like cholestasis and phospholipidosis [56]. Kemboi et al. [57] discovered that other developing toxins and metabolites, counting *Alternaria, Aspergillus, Fusarium, Penicillium* toxins, were discovered at differing concentrations during their research. Such co-occurrences of mycotoxins could trigger synergistic and additive health effects, impeding the food animal production sectors worldwide.

Results of feed consumption and feed conversion ratio are shown in Tables 4 and 5. *Alternaria* spp. contaminated wheat grain showed some numerical differences between treatments but without any statistically significant differences in broiler chickens' life stage of life.

Table 4. Feed consumption of broiler chickens, g.

Age	Treatments in Test			Pooled	
	A1	**A2**	**A3**	**SE**	*p*
7 day	163.57 [a] ± 30.51	152.63 [a] ± 20.01	162.71 [a] ± 29.50	9.62	0.564
14 day	292.33 [a] ± 10.16	293.77 [a] ± 11.66	284.11 [a] ± 17.03	4.72	0.689

Values in the same row marked with the different letters are significantly different at the significance level of $p < 0.05$.

Table 5. The feed conversion ratio of broiler chickens, kg/kg.

Age	Treatments in Test			Pooled	
	A1	**A2**	**A3**	**SE**	*p*
7 day	1.16 [a] ± 0.21	1.11 [a] ± 0.15	1.21 [a] ± 0.22	0.07	0.454
14 day	1.29 [a] ± 0.09	1.29 [a] ± 0.07	1.33 [a] ± 0.08	0.04	0.555

Values in the same row marked with the different letters are significantly different at the significance level of $p < 0.05$.

In addition to wheat, corn is the main feed ingredient used in poultry nutrition. As a wheat grain, the corn can also be naturally infected with mycotoxins, especially with *Alternaria* spp. Topi et al. [58] have investigated the presence of *Alternaria* mycotoxins in grains from Albania: alternariol, alternariol monomethyl ether, tenuazonic acid, and tentoxin. They have concluded that the contribution of AOH and AME originating from wheat was 0–31.7 ng/kg body weight per day. In contrast, the contribution of *Alternaria* toxins through maize consumption was significantly lower.

Changes from optimal internal and external conditions lead to stress from a physiological point of view. Between the main stressors in broiler production, nutritional stressors have a significant role, and within them, the leading role is mycotoxins feed contamination [34].

The highly probable clarification for the remarked results presented in Table 6 is that the pathological modifications strengthen free radical processes by promoting catalytic activities of enzymes engaged in the antioxidative protection, POD, and GR. Still, through the disease phase, lipolysis from the lipid depots could be increased due to reduced feed consumption, which is not the case in our research. Moreover, tiredness of the organism could lead to escalation of free radical processes and higher amounts of lipid peroxides in blood. To defend himself, the body initiates its antioxidative safety mechanisms. Decrease of SOD activity was anticipated and is in accordance with other research [59,60].

Table 6. GSH and LPx content and the activity of POD, SOD, GR, and XOD in blood hemolysates, µmol/g Hb min.

Treatment	GSH	LPx	POD	SOD	GR	XOD
A1	5.1 [b] ± 0.9	0.3 [a] ± 0.2	65.2 [a] ± 4.4	82.6 [a] ± 6.1	11.3 [b] ± 5.9	26.3 [a] ± 3.6
A2	5.8 [a] ± 0.3	0.3 [a] ± 0.1	55.1 [b] ± 7.2	25.5 [b] ± 3.5	21.1 [a] ± 7.2	27.5 [a] ± 2.9
A3	6.2 [a] ± 1.1	0.4 [a] ± 0.1	59.3 [b] ± 3.4	29.1 [b] ± 8.9	19.8 [a] ± 9.6	26.7 [a] ± 4.1
p-value	0.032	0.089	0.038	0.001	0.004	0.341

Values in the same row marked with the different letters are significantly different at the significance level of $p < 0.05$.

The glutathione has a vital position in reducing the acute toxicity of xenobiotics and products of lipid peroxidation. A statistically significant decrease of POD activity compared to the A1 control treatment was expected since POD catalyzes various proton donors' oxidation with hydrogen peroxide. Having in mind that mycotoxins are classified as hepatotoxins, nephrotoxins, neurotoxins, immunotoxins, and that there are to date, 400 mycotoxins identified and the most critical species producing mycotoxins belong to *Aspergillus, Penicillium, Alternaria*, and *Fusarium* genera, Ülger et al. [61] have described their genotoxic effects on the organism. Uric acid increased accumulation, and reduced excretion is closely related to the pathogenesis of gout and hyperuricemia. Higher plants produce different metabolites, which might impede XOD, so disallow the oxidation of hypoxanthine to xanthine then to uric acid in the purine metabolism. Nevertheless, microorganisms generate a group of degrading enzymes, which catalyze uric acid degradation to ammonia. Xanthine oxidoreductase (XOR) has two forms; xanthine oxidase (XOD) and xanthine dehydrogenase (XDH), both of them catalyze the oxidation of hypoxanthine to xanthines, then to uric acid in the purine metabolism [62]. Hafez et al. [63] presented an analysis with the incidence of uric acid in plants and other organisms, especially microorganisms, in addition to the mechanisms by which plant extracts, metabolites, and enzymes could reduce uric acid in the blood. Overactivity of both enzymes (XOD and XDH) cause the accumulation of uric acid in the animal body and form a pathogenesis condition called gout [64]. Additionally, XOD serves as a valuable biological source of oxygen free radicals that participate in various damages of animal tissues leading to many pathological states [65], which could be caused by multiple stress triggers, e.g., mycotoxins [66–68].

Serum biochemical parameters were significantly affected by *Alternaria* spp. wheat in both treatments compared to control treatment during the starter dietary phase (Table 7). Even though the *Alternaria* spp. contaminated wheat had no significant effect on growth performance in broiler chicks, it induced the typical clinical signs of hepatic injury, including increased activities of AST and ALT, during the starter dietary period what is in accordance with results of other researchers [48,69,70].

Table 7. Aminotransferase (AST) and alanine aminotransferase (ALT) activity in serum of broiler chickens, U/L.

Treatment	AST	ALT
A1	182.1 [b] ± 32.1	1.1 [b] ± 0.1
A2	268.8 [a] ± 41.3	1.7 [a] ± 0.3
A3	271.5 [a] ± 33.7	1.6 [a] ± 0.2
p-value	0.075	0.039

Values in the same row marked with the different letters are significantly different at the significance level of $p < 0.05$.

Oxidative stress plays an important role in the development of many animal diseases and it has been shown that have significant implications for the well-being and overall welfare of nonruminants [71]. Various studies have shown that oxidative stress has a fundamental role in the etiopathogenesis of several acute and chronic diseases which are causally related to animal welfare [72]. Over the years oxidative stress has been deeply investigated in human, while in poultry production the data are yet less uneven [73]. Poultry welfare is fundamental in maintaining correct health and a good level of mental and physical well-being of the animal [74]. In our study increased content of total glutathione levels in chicken dietary treatments (5.8 and 6.2 µmol/g Hb min) with addition of blackpoint wheat, indicates that chickens had increased antioxidant defense. These results are directly related with the impaired welfare of chickens. Likewise, certain indicators of impaired welfare of chickens in our expert are increased activity of GR (21.1 and 19.8 µmol/g Hb min), and decreased activity of SOD (25.5 and 29.1 µmol/g Hb min), respectively. The similar results were obtained by Brambilla et al. [75] in their research related to influence of oxidative stress markers reactive oxygen metabolites (ROM) and anti-oxidant power (OXY) in swine welfare. Stresses in commercial poultry result from many various factors which negatively impact poultry health, production, and welfare [76]. Oxidative stress is downstream of all these stresses. Oxidative stress in the cells results from an imbalance between free radical production and endogenous antioxidant defense [77]. It is well documented that poultry feed is often contaminated with a wide range of environmental toxicants, bacterial and fungal toxins, and known to affect the health and welfare of poultry [78]. Mycotoxins usually generates reactive oxygen species which induces lipid peroxidation, alters the cellular redox signaling, antioxidant status, and membrane integrity of the cells [79]. Mycotoxins increase cellular apoptosis and affect poultry health, production, and welfare.

4. Conclusions

Based on the gained results, it can be concluded that identifying field fungi in all analyzed wheat grain samples showed that the dominant mycotoxigenic fungus was *Rhizopus* spp., followed by *Alternaria* spp., and *Fusarium* spp. Usage of 25% wheat in complete feed for broiler chickens in the first 14 days of life has shown adverse effects reflected on body weight gain without significant influence on feed consumption and utilization. Concerning oxidative stress, it can be concluded that *Alternaria* spp. causes high oxidative stress in chickens at a young age negatively influences production and overall broiler chicken's welfare.

Further research on the influence of *Alternaria* spp. on animal production and genotoxicity is still essential.

Author Contributions: Conceptualization, N.P.; methodology, V.B. and S.T.; software, N.P. and D.M.; validation, M.C., G.V., J.M. and T.S.K.; formal analysis, N.P., V.B., S.T. and G.V.; investigation, N.P.; resources, N.P.; data curation, A.P.; writing—original draft preparation, N.P.; writing—review and editing, V.B. and M.C.; visualization, N.P.; supervision, M.C., S.T., G.V. and T.S.K.; project administration, N.P.; funding acquisition, N.P. and V.B. All authors have read and agreed to the published version of the manuscript.

Funding: This research was funded by the Ministry for Education, Science and Technological Development of the Republic of Serbia.

Institutional Review Board Statement: The study was conducted according to the guidelines of the Declaration of Helsinki and approved by the Ethics Committee of University Business Academy in Novi Sad, EC-2020-321/9, 13.03.2020.

Informed Consent Statement: Not applicable.

Data Availability Statement: Data is contained within the article.

Acknowledgments: This research was supported by the Ministry for Education, Science and Technological Development of the Republic of Serbia within the postdoctoral scholarship in Toxicology and Molecular Genetics under the Grant 451-03-1002/2020-14.

Conflicts of Interest: The authors declare no conflict of interest.

References

1. Beloshapka, A.; Buff, P.; Fahey, G.; Swanson, K. Compositional Analysis of Whole Grains, Processed Grains, Grain Co-Products, and Other Carbohydrate Sources with Applicability to Pet Animal Nutrition. *Foods* **2016**, *5*, 23. [CrossRef] [PubMed]
2. Fraeyman, S.; Croubels, S.; Devreese, M.; Antonissen, G. Emerging Fusarium and Alternaria Mycotoxins: Occurrence, Toxicity and Toxicokinetics. *Toxins* **2017**, *9*, 228. [CrossRef]
3. Thomma, B.P.H.J. *Alternaria* Spp.: From General Saprophyte to Specific Parasite. *Alternaria Mol. Plant Pathol.* **2003**, *4*, 225–236. [CrossRef] [PubMed]
4. Kusaba, M.; Tsuge, T. Phologeny of Alternaria Fungi Known to Produce Host-Specific Toxins on the Basis of Variation in Internal Transcribed Spacers of Ribosomal DNA. *Curr. Genet.* **1995**, *28*, 491–498. [CrossRef]
5. Puvača, N.; Bursić, V.; Vuković, G.; Budakov, D.; Petrović, A.; Merkuri, J.; Avantaggiato, G.; Cara, M. Ascomycete Fungi (Alternaria Spp.) Characterization as Major Feed Grains Pathogens. *J. Agron. Technol. Eng. Manag.* **2020**, *3*, 499–505.
6. Hillmann, F.; Novohradská, S.; Mattern, D.J.; Forberger, T.; Heinekamp, T.; Westermann, M.; Winckler, T.; Brakhage, A.A. Virulence Determinants of the Human Pathogenic Fungus *A Spergillus Fumigatus* Protect against Soil Amoeba Predation: *Dictyostelium* Interactions with *Aspergillus Fumigatus*. *Environ. Microbiol.* **2015**, *17*, 2858–2869. [CrossRef] [PubMed]
7. Ferreira, R.B.; Monteiro, S.; Freitas, R.; Santos, C.N.; Chen, Z.; Batista, L.M.; Duarte, J.; Borges, A.; Teixeira, A.R. Fungal Pathogens: The Battle for Plant Infection. *Crit. Rev. Plant Sci.* **2006**, *25*, 505–524. [CrossRef]
8. Jastrzębska, M.; Wachowska, U.; Kostrzewska, M.K. Pathogenic and Non-Pathogenic Fungal Communities in Wheat Grain as Influenced by Recycled Phosphorus Fertilizers: A Case Study. *Agriculture* **2020**, *10*, 239. [CrossRef]
9. Müller, M.E.H.; Urban, K.; Köppen, R.; Siegel, D.; Korn, U.; Koch, M. Mycotoxins as Antagonistic or Supporting Agents in the Interaction between Phytopathogenic *Fusarium* and *Alternaria* Fungi. *World Mycotoxin J.* **2015**, *8*, 311–321. [CrossRef]
10. Li, Q.-Y.; Xu, Q.-Q.; Jiang, Y.-M.; Niu, J.-S.; Xu, K.-G.; He, R.-S. The Correlation between Wheat Black Point and Agronomic Traits in the North China Plain. *Crop. Prot.* **2019**, *119*, 17–23. [CrossRef]
11. Lv, G.; Dong, Z.; Wang, Y.; Geng, J.; Li, J.; Lv, X.; Sun, C.; Ren, Y.; Zhang, J.; Chen, F. Identification of Genetic Loci of Black Point in Chinese Common Wheat by Genome-Wide Association Study and Linkage Mapping. *Plant Dis.* **2020**, *104*, 2005–2013. [CrossRef]
12. Puvača, N.; Bursić, V.; Petrović, A.; Vuković, G.; Cara, M.; Peulić, T.; Avantaggiato, G. Mycotoxin Incidence of Ochratoxin A in Wine and Methods for Its Control. *J. Agron. Technol. Eng. Manag.* **2020**, *3*, 475–482.
13. Čolović, R.; Puvača, N.; Cheli, F.; Avantaggiato, G.; Greco, D.; Đuragić, O.; Kos, J.; Pinotti, L. Decontamination of Mycotoxin-Contaminated Feedstuffs and Compound Feed. *Toxins* **2019**, *11*, 617. [CrossRef] [PubMed]
14. Logrieco, A.; Bottalico, A.; Mulé, G.; Moretti, A.; Perrone, G. Epidemiology of toxigenic fungi and their associated mycotoxins for some Mediterranean crops. In *Epidemiology of Mycotoxin Producing Fungi: Under the Aegis of COST Action 835 'Agriculturally Important Toxigenic Fungi 1998–2003', EU Project (QLK 1-CT-1998–01380)*; Xu, X., Bailey, J.A., Cooke, B.M., Eds.; Springer: Dordrecht, The Netherlands, 2003; pp. 645–667. ISBN 978-94-017-1452-5.
15. Somma, S.; Amatulli, M.T.; Masiello, M.; Moretti, A.; Logrieco, A.F. Alternaria Species Associated to Wheat Black Point Identified through a Multilocus Sequence Approach. *Int. J. Food Microbiol.* **2019**, *293*, 34–43. [CrossRef]
16. da Cruz Cabral, L.; Terminiello, L.; Fernández Pinto, V.; Fog Nielsen, K.; Patriarca, A. Natural Occurrence of Mycotoxins and Toxigenic Capacity of Alternaria Strains from Mouldy Peppers. *Int. J. Food Microbiol.* **2016**, *236*, 155–160. [CrossRef] [PubMed]
17. da Cruz Cabral, L.; Delgado, J.; Patriarca, A.; Rodríguez, A. Differential Response to Synthetic and Natural Antifungals by Alternaria Tenuissima in Wheat Simulating Media: Growth, Mycotoxin Production and Expression of a Gene Related to Cell Wall Integrity. *Int. J. Food Microbiol.* **2019**, *292*, 48–55. [CrossRef]
18. Vejdovszky, K.; Sack, M.; Jarolim, K.; Aichinger, G.; Somoza, M.M.; Marko, D. In Vitro Combinatory Effects of the Alternaria Mycotoxins Alternariol and Altertoxin II and Potentially Involved MiRNAs. *Toxicol. Lett.* **2017**, *267*, 45–52. [CrossRef]
19. López, P.; Venema, D.; de Rijk, T.; de Kok, A.; Scholten, J.M.; Mol, H.G.J.; de Nijs, M. Occurrence of Alternaria Toxins in Food Products in The Netherlands. *Food Control* **2016**, *60*, 196–204. [CrossRef]
20. Nawaz, S.; Scudamore, K.A.; Rainbird, S.C. Mycotoxins in Ingredients of Animal Feeding Stuffs: I. Determination of *Alternaria* Mycotoxins in Oilseed Rape Meal and Sunflower Seed Meal. *Food Addit. Contam.* **1997**, *14*, 249–262. [CrossRef]
21. Escrivá, L.; Oueslati, S.; Font, G.; Manyes, L. *Alternaria* Mycotoxins in Food and Feed: An Overview. *J. Food Qual.* **2017**, *2017*, 1–20. [CrossRef]
22. Puntscher, H.; Aichinger, G.; Grabher, S.; Attakpah, E.; Krüger, F.; Tillmann, K.; Motschnig, T.; Hohenbichler, J.; Braun, D.; Plasenzotti, R.; et al. Bioavailability, Metabolism, and Excretion of a Complex Alternaria Culture Extract versus Altertoxin II: A Comparative Study in Rats. *Arch. Toxicol.* **2019**, *93*, 3153–3167. [CrossRef]
23. Ostry, V. *Alternaria* Mycotoxins: An Overview of Chemical Characterization, Producers, Toxicity, Analysis and Occurrence in Foodstuffs. *World Mycotoxin J.* **2008**, *1*, 175–188. [CrossRef]
24. Crudo, F.; Aichinger, G.; Mihajlovic, J.; Dellafiora, L.; Varga, E.; Puntscher, H.; Warth, B.; Dall'Asta, C.; Berry, D.; Marko, D. Gut Microbiota and Undigested Food Constituents Modify Toxin Composition and Suppress the Genotoxicity of a Naturally Occurring Mixture of Alternaria Toxins in Vitro. *Arch. Toxicol.* **2020**, *94*, 3541–3552. [CrossRef]
25. Meena, M.; Swapnil, P.; Upadhyay, R.S. Isolation, Characterization and Toxicological Potential of Alternaria-Mycotoxins (TeA, AOH and AME) in Different Alternaria Species from Various Regions of India. *Sci. Rep.* **2017**, *7*, 8777. [CrossRef] [PubMed]
26. Patriarca, A.; Fernández Pinto, V. Prevalence of Mycotoxins in Foods and Decontamination. *Curr. Opin. Food Sci.* **2017**, *14*, 50–60. [CrossRef]

27. Ramires, F.; Masiello, M.; Somma, S.; Villani, A.; Susca, A.; Logrieco, A.; Luz, C.; Meca, G.; Moretti, A. Phylogeny and Mycotoxin Characterization of Alternaria Species Isolated from Wheat Grown in Tuscany, Italy. *Toxins* **2018**, *10*, 472. [CrossRef]
28. Gotthardt, M.; Asam, S.; Gunkel, K.; Moghaddam, A.F.; Baumann, E.; Kietz, R.; Rychlik, M. Quantitation of Six Alternaria Toxins in Infant Foods Applying Stable Isotope Labeled Standards. *Front. Microbiol.* **2019**, *10*, 109. [CrossRef]
29. Fremy, J.-M.; Alassane-Kpembi, I.; Oswald, I.P.; Cottrill, B.; Van Egmond, H.P. A Review on Combined Effects of Moniliformin and Co-Occurring *Fusarium* Toxins in Farm Animals. *World Mycotoxin J.* **2019**, *12*, 281–291. [CrossRef]
30. Puvača, N.; Bursić, V.; Petrović, A.; Prodanović, R.; Mansour, M.K.; Obućinski, D.; Vuković, G.; Marić, M. Influence of Tea Tree Essential Oil on the Synthesis of Mycotoxins: Ochratoxin A. *Maced. J. Anim. Sci.* **2019**, *9*, 25–29.
31. Puvača, N.; Ljubojevic, D.; Živkov Baloš, M.; Đuragić, O.; Bursić, V.; Vuković, G.; Prodanović, R.; Bošković, J. Occurance of Mycotoxins and Mycotoxicosis in Poultry. *CDVS* **2018**, *2*, 165–167. [CrossRef]
32. Saleh, K.M.M.; Al-Zghoul, M.B. Effect of Acute Heat Stress on the MRNA Levels of Cytokines in Broiler Chickens Subjected to Embryonic Thermal Manipulation. *Animals* **2019**, *9*, 499. [CrossRef] [PubMed]
33. Hasan, R.; Lasker, S.; Hasan, A.; Zerin, F.; Zamila, M.; Chowdhury, F.I.; Nayan, S.I.; Rahman, M.M.; Khan, F.; Subhan, N.; et al. Canagliflozin Attenuates Isoprenaline-Induced Cardiac Oxidative Stress by Stimulating Multiple Antioxidant and Anti-Inflammatory Signaling Pathways. *Sci. Rep.* **2020**, *10*, 14459. [CrossRef] [PubMed]
34. Surai, P.F.; Kochish, I.I.; Fisinin, V.I.; Kidd, M.T. Kidd Antioxidant Defence Systems and Oxidative Stress in Poultry Biology: An Update. *Antioxidants* **2019**, *8*, 235. [CrossRef]
35. Abuelo, A.; Hernández, J.; Benedito, J.L.; Castillo, C. Redox Biology in Transition Periods of Dairy Cattle: Role in the Health of Periparturient and Neonatal Animals. *Antioxidants* **2019**, *8*, 20. [CrossRef] [PubMed]
36. Preti, R.; Tarola, A.M. Study of Polyphenols, Antioxidant Capacity and Minerals for the Valorisation of Ancient Apple Cultivars from Northeast Italy. *Eur. Food Res. Technol.* **2020**, *247*, 273–283. [CrossRef]
37. Surai, P.F.; Kochish, I.I. Antioxidant Systems and Vitagenes in Poultry Biology: Heat Shock Proteins. In *Heat Shock Proteins in Veterinary Medicine and Sciences*; Heat Shock Proteins; Asea, A.A.A., Kaur, P., Eds.; Springer International Publishing: Cham, Switzerland, 2017; Volume 12, pp. 123–177. ISBN 978-3-319-73376-0.
38. Liguori, I.; Russo, G.; Curcio, F.; Bulli, G.; Aran, L.; Della-Morte, D.; Gargiulo, G.; Testa, G.; Cacciatore, F.; Bonaduce, D.; et al. Oxidative Stress, Aging, and Diseases. *Clin. Interv. Aging* **2018**, *13*, 757–772. [CrossRef]
39. Gessner, D.K.; Ringseis, R.; Eder, K. Potential of Plant Polyphenols to Combat Oxidative Stress and Inflammatory Processes in Farm Animals. *J. Anim. Physiol. Anim. Nutr.* **2017**, *101*, 605–628. [CrossRef]
40. Kostadinović, L.; Lević, J. Effects of Phytoadditives in Poultry and Pigs Diseases. *J. Agron. Technol. Eng. Manag.* **2018**, *1*, 1–7.
41. Layé, S. Polyunsaturated Fatty Acids, Neuroinflammation and Well Being. *Prostaglandins Leukot. Essential Fat. Acids (PLEFA)* **2010**, *82*, 295–303. [CrossRef] [PubMed]
42. AOAC. *Official Methods of Analysis*, 16th ed.; Association of Analytical Communities: Gaithersburg, MD, USA, 1998.
43. Chae, Y.; Hwang, J. Visual Color Difference between Colored-Yarn Mixed Woven Fabrics and Their Instrumentally Measured Colors: The Effects of Individual Yarn Colors and Texture. *Fibers Polym.* **2020**, *21*, 792–802. [CrossRef]
44. Pitt, J.I.; Hocking, A.D. *Fungi and Food Spoilage*; Springer: New York, NY, USA, 2009.
45. Kostadinović, L.M.; Popović, S.J.; Puvača, N.M.; Čabarkapa, I.S.; Kormanjoš, Š.M.; Lević, J.D. Influence of Artemisia Absinthium Essential Oil on Antioxidative System of Broilers Experimentally Infected with Eimeria Oocysts. *Vet. Arh.* **2016**, *86*, 253–264.
46. Popović, S.; Puvača, N.; Kostadinović, L.; Džinić, N.; Bošnjak, J.; Vasiljević, M.; Djuragic, O. Effects of Dietary Essential Oils on Productive Performance, Blood Lipid Profile, Enzyme Activity and Immunological Response of Broiler Chickens. *Eur. Poult. Sci.* **2016**, *80*, 146. [CrossRef]
47. Puvača, N.; Kostadinović, L.; Popović, S.; Lević, J.; Ljubojević, D.; Tufarelli, V.; Jovanović, R.; Tasić, T.; Ikonić, P.; Lukač, D. Proximate Composition, Cholesterol Concentration and Lipid Oxidation of Meat from Chickens Fed Dietary Spice Addition (Allium Sativum, Piper Nigrum, Capsicum Annuum). *Anim. Prod. Sci.* **2016**, *56*, 1920–1927. [CrossRef]
48. Zhang, N.-Y.; Qi, M.; Zhao, L.; Zhu, M.-K.; Guo, J.; Liu, J.; Gu, C.-Q.; Rajput, S.; Krumm, C.; Qi, D.-S.; et al. Curcumin Prevents Aflatoxin B1 Hepatoxicity by Inhibition of Cytochrome P450 Isozymes in Chick Liver. *Toxins* **2016**, *8*, 327. [CrossRef]
49. Funnell, D.L.; Pedersen, J.F. Association of Plant Color and Pericarp Color with Colonization of Grain by Members of *Fusarium* and *Alternaria* in Near-Isogenic Sorghum Lines. *Plant Dis.* **2006**, *90*, 411–418. [CrossRef]
50. Janic-Hajnal, E.; Belovic, M.; Plavsic, D.; Mastilovic, J.; Bagi, F.; Budakov, D.; Kos, J. Visual, Instrumental, Mycological and Mycotoxicological Characterization of Wheat Inoculated with and Protected against Alternaria Spp. *Hem. Ind.* **2016**, *70*, 257–264. [CrossRef]
51. Xu, W.; Han, X.; Li, F.; Zhang, L. Natural Occurrence of Alternaria Toxins in the 2015 Wheat from Anhui Province, China. *Toxins* **2016**, *8*, 308. [CrossRef] [PubMed]
52. Li, F.; Yoshizawa, T. *Alternaria* Mycotoxins in Weathered Wheat from China. *J. Agric. Food Chem.* **2000**, *48*, 2920–2924. [CrossRef] [PubMed]
53. Kosiak, B.; Torp, M.; Skjerve, E.; Andersen, B. Alternaria and Fusarium in Norwegian Grains of Reduced Quality—A Matched Pair Sample Study. *Int. J. Food Microbiol.* **2004**, *93*, 51–62. [CrossRef]
54. Giambrone, J.J.; Davis, N.D.; Diener, U.L. Effect of Tenuazonic Acid on Young Chickens. *Poult. Sci.* **1978**, *57*, 1554–1558. [CrossRef]

55. Kolawole, O.; Graham, A.; Donaldson, C.; Owens, B.; Abia, W.A.; Meneely, J.; Alcorn, M.J.; Connolly, L.; Elliott, C.T. Low Doses of Mycotoxin Mixtures below EU Regulatory Limits Can Negatively Affect the Performance of Broiler Chickens: A Longitudinal Study. *Toxins* **2020**, *12*, 433. [CrossRef]

56. Hessel-Pras, S.; Kieshauer, J.; Roenn, G.; Luckert, C.; Braeuning, A.; Lampen, A. In Vitro Characterization of Hepatic Toxicity of Alternaria Toxins. *Mycotoxin Res.* **2019**, *35*, 157–168. [CrossRef]

57. Kemboi, D.C.; Ochieng, P.E.; Antonissen, G.; Croubels, S.; Scippo, M.-L.; Okoth, S.; Kangethe, E.K.; Faas, J.; Doupovec, B.; Lindahl, J.F.; et al. Multi-Mycotoxin Occurrence in Dairy Cattle and Poultry Feeds and Feed Ingredients from Machakos Town, Kenya. *Toxins* **2020**, *12*, 762. [CrossRef]

58. Topi, D.; Tavčar-Kalcher, G.; Pavšič-Vrtač, K.; Babič, J.; Jakovac-Strajn, B. *Alternaria* Mycotoxins in Grains from Albania: Alternariol, Alternariol Monomethyl Ether, Tenuazonic Acid and Tentoxin. *World Mycotoxin J.* **2019**, *12*, 89–99. [CrossRef]

59. Kustrzeba-Wójcicka, I.; Siwak, E.; Terlecki, G.; Wolańczyk-Mędrala, A.; Mędrala, W. Alternaria Alternata and Its Allergens: A Comprehensive Review. *Clin. Rev. Allerg. Immunol.* **2014**, *47*, 354–365. [CrossRef] [PubMed]

60. AlMatar, M.; Var, I.; Koksal, F. How Does Alternaria Alternata-Derived Alternariol Affect Our Health? *Mini Rev. Org. Chem.* **2016**, *13*, 465–472. [CrossRef]

61. Ülger, T.G.; Uçar, A.; Çakıroğlu, F.P.; Yilmaz, S. Genotoxic Effects of Mycotoxins. *Toxicon* **2020**, *185*, 104–113. [CrossRef] [PubMed]

62. Unno, T.; Sugimoto, A.; Kakuda, T. Xanthine Oxidase Inhibitors from the Leaves of Lagerstroemia Speciosa (L.) Pers. *J. Ethnopharmacol.* **2004**, *93*, 391–395. [CrossRef]

63. Hafez, R.M.; Abdel-Rahman, T.M.; Naguib, R.M. Uric Acid in Plants and Microorganisms: Biological Applications and Genetics—A Review. *J. Adv. Res.* **2017**, *8*, 475–486. [CrossRef]

64. Lioté, F. Hyperuricemia and Gout. *Curr. Rheumatol. Rep.* **2003**, *5*, 227–234. [CrossRef]

65. Alsultanee, I.R.; Ewadh, M.J.; Mohammed, M.F. Novel Natural Anti Gout Medication Extract from Momordica Charantia. *J. Nat. Sci. Res.* **2014**, *4*, 16–23.

66. Hou, Y.-J.; Zhao, Y.-Y.; Xiong, B.; Cui, X.-S.; Kim, N.-H.; Xu, Y.-X.; Sun, S.-C. Mycotoxin-Containing Diet Causes Oxidative Stress in the Mouse. *PLoS ONE* **2013**, *8*, e60374. [CrossRef]

67. Richard, J.L. Some Major Mycotoxins and Their Mycotoxicoses—An Overview. *Int. J. Food Microbiol.* **2007**, *119*, 3–10. [CrossRef] [PubMed]

68. Doi, K.; Uetsuka, K. Mechanisms of Mycotoxin-Induced Neurotoxicity through Oxidative Stress-Associated Pathways. *Int. J. Mol. Sci.* **2011**, *12*, 5213–5237. [CrossRef] [PubMed]

69. Sun, L.-H.; Zhang, N.-Y.; Sun, R.-R.; Gao, X.; Gu, C.; Krumm, C.S.; Qi, D.-S. A Novel Strain of Cellulosimicrobium Funkei Can Biologically Detoxify Aflatoxin B1 in Ducklings. *Microb. Biotechnol.* **2015**, *8*, 490–498. [CrossRef]

70. Klein, P.J.; Van Vleet, T.R.; Hall, J.O.; Coulombe, R.A., Jr. Biochemical Factors Underlying the Age-Related Sensitivity of Turkeys to Aflatoxin B1. *Comp. Biochem. Physiol. Part C Toxicol. Pharmacol.* **2002**, *132*, 193–201. [CrossRef]

71. Molinari, L.; Basini, G.; Ramoni, R.; Bussolati, S.; Aldigeri, R.; Grolli, S.; Bertini, S.; Quintavalla, F. Evaluation of Oxidative Stress Parameters in Healthy Saddle Horses in Relation to Housing Conditions, Presence of Stereotypies, Age, Sex and Breed. *Processes* **2020**, *8*, 1670. [CrossRef]

72. Reiser, S.; Wuertz, S.; Schroeder, J.P.; Kloas, W.; Hanel, R. Risks of Seawater Ozonation in Recirculation Aquaculture—Effects of Oxidative Stress on Animal Welfare of Juvenile Turbot (Psetta Maxima, L.). *Aquat. Toxicol.* **2011**, *105*, 508–517. [CrossRef]

73. Altan, Ö.; Pabuçcuoğlu, A.; Altan, A.; Konyalioğlu, S.; Bayraktar, H. Effect of Heat Stress on Oxidative Stress, Lipid Peroxidation and Some Stress Parameters in Broilers. *Br. Poult. Sci.* **2003**, *44*, 545–550. [CrossRef] [PubMed]

74. Carenzi, C.; Verga, M. Animal Welfare: Review of the Scientific Concept and Definition. *Ital. J. Anim. Sci.* **2009**, *8*, 21–30. [CrossRef]

75. Brambilla, G.; Civitareale, C.; Ballerini, A.; Fiori, M.; Amadori, M.; Archetti, L.I.; Regini, M.; Betti, M. Response to Oxidative Stress as a Welfare Parameter in Swine. *Redox Rep.* **2002**, *7*, 159–163. [CrossRef]

76. Wasti, S.; Sah, N.; Mishra, B. Impact of Heat Stress on Poultry Health and Performances, and Potential Mitigation Strategies. *Animals* **2020**, *10*, 1266. [CrossRef] [PubMed]

77. Shankar, K.; Mehendale, H.M. Oxidative Stress. In *Encyclopedia of Toxicology*; Elsevier: Amsterdam, The Netherlands, 2014; pp. 735–737, ISBN 978-0-12-386455-0.

78. Murugesan, G.R.; Ledoux, D.R.; Naehrer, K.; Berthiller, F.; Applegate, T.J.; Grenier, B.; Phillips, T.D.; Schatzmayr, G. Prevalence and Effects of Mycotoxins on Poultry Health and Performance, and Recent Development in Mycotoxin Counteracting Strategies. *Poult. Sci.* **2015**, *94*, 1298–1315. [CrossRef] [PubMed]

79. Mishra, B.; Jha, R. Oxidative Stress in the Poultry Gut: Potential Challenges and Interventions. *Front. Vet. Sci* **2019**, *6*, 60. [CrossRef] [PubMed]

 *sustainability*

Article

Effects of Sunflower Meal Supplementation as a Complementary Protein Source in the Laying Hen's Diet on Productive Performance, Egg Quality, and Nutrient Digestibility

Ahmed A. Saleh [1],[*], Ahmed El-Awady [1], Khairy Amber [1], Yahya Z. Eid [1], Mohammed H. Alzawqari [1],[2], Shaimaa Selim [3], Mohamed Mohamed Soliman [4] and Mustafa Shukry [5],[*]

[1] Department of Poultry Production, Faculty of Agriculture, Kafrelsheikh University, Kafrelsheikh 333516, Egypt; elawody@gmail.com (A.E.-A.); khairyamber1957@gmail.com (K.A.); yahya.eid@agr.kfs.edu.eg (Y.Z.E.); m.alzawqari@gmail.com (M.H.A.)

[2] Department of Animal Production, Faculty of Agriculture and Veterinary Medicine, Ibb University, Ibb 70270, Yemen

[3] Department of Nutrition and Clinical Nutrition, Faculty of Veterinary Medicine, Menoufia University, Shibin El-Kom 32514, Egypt; shimaaselim@vet.menofia.edu.eg

[4] Clinical Laboratory Sciences Department, Turabah University College, Taif University, P.O. Box 11099, Taif 21944, Saudi Arabia; mmsoliman@tu.edu.sa

[5] Department of Physiology, Faculty of Veterinary Medicine, Kafrelsheikh University, Kafrelsheikh 33511, Egypt

[*] Correspondence: ahmed.saleh1@agr.kfs.edu.eg (A.A.S.); mostafa.ataa@vet.kfs.edu.eg (M.S.)

check for **updates**

Citation: Saleh, A.A.; El-Awady, A.; Amber, K.; Eid, Y.Z.; Alzawqari, M.H.; Selim, S.; Soliman, M.M.; Shukry, M. Effects of Sunflower Meal Supplementation as a Complementary Protein Source in the Laying Hen's Diet on Productive Performance, Egg Quality, and Nutrient Digestibility. *Sustainability* **2021**, *13*, 3557. https://doi.org/10.3390/su13063557

Academic Editors: Marc A. Rosen and Nikola Puvača

Received: 14 February 2021
Accepted: 20 March 2021
Published: 23 March 2021

Publisher's Note: MDPI stays neutral with regard to jurisdictional claims in published maps and institutional affiliations.

Abstract: The practical usage of untraditional feedstuffs such as sunflower meal (SFM) in laying hens nutrition in developing countries has received considerable attention. SFM is a by-product of the sunflower oil industry and has been progressively added to bird's diets. Sunflower meal (SFM) is gaining great interest as a feed ingredient due to its eminent crude protein content, low anti-nutritional compounds, and low price. The current experiment was aimed to assess the production efficiency, egg quality, yolk fatty acids composition, and nutrient digestibility of laying hens fed SFM. A total of 162 Bovans Brown laying hens aged 60 weeks old were randomly allocated using a completely randomized design into three experimental groups of nine replicates each (n = six/replicate) for eight weeks. The dietary treatments involved a control (basal diet) and two levels of SFM, 50 and 100 g/kg feed. The dietary treatments did not influence live weight gain, feed intake, and egg mass. On one hand, the laying rate was increased; on the other hand, the feed conversion ratio and broken eggs rate of laying hens were decreased ($p < 0.05$) by the dietary inclusion of SFM. Dietary treatments had no effect on the egg's quality characteristics except the yolk color and yolk height were larger ($p = 0.01$) for laying hens fed SFM compared with those fed the control. Dietary inclusion of SFM decreased ($p < 0.05$) the content of cholesterol in the egg yolk. Still, it increased the yolk contents of vitamin E, calcium, linoleic acid, linolenic acid, and oleic acid ($p < 0.05$). Furthermore, the dietary inclusion of SFM increased crude protein and calcium digestibility, but decreased the ether extract digestibility. In conclusion, our results suggested that the dietary inclusion of SFM, up to 100 g/kg at a late phase of laying, could improve the production performance, some of the egg quality traits, and nutrient digestibility while decreasing egg yolk cholesterol.

Keywords: sunflower meal; laying hens; performance; yolk cholesterol; yolk fatty acids; egg quality

1. Introduction

In the practical poultry industry, higher feed ingredient prices have led to a closer consideration to seek less expensive agricultural by-products [1]. Sources of protein are becoming more and more limited around the world. Consequently, there is a need to search for alternative protein sources [2]. In general, feed costs reflect much of the expenses, and abrupt increases in feed costs make it a contest for nutritionists to sustain animal production and safety while balancing the cost of the diet [3]. Soybean meal (SBM) is one of the most popular sources of protein used in poultry diets. When the price rises, nutritionists must choose the available ingredients to formulate cheap, balanced, and economically viable diets [4,5]. In this context, developing diet formulations with alternative ingredients is the best to overcome this problem and reduce feed costs, especially when these alternative ingredients are locally available.

Sunflower can be harvested in tropical areas two or three times a year, and it is a healthy substitute for the oil manufacturers and the feed mill district [6]. Sunflower meal (SFM) is an invention from the oil extraction of sunflower seed, and it is utilized primarily as protein and fiber sources in the diets of poultry [4,5,7–9]. Although SFM is opulent in crude protein, its poultry applications have some limitations due to its relatively extreme fiber including insoluble fiber and low levels of specific limiting amino acids such as lysine and methionine. Additionally, sunflower seeds have a high content in α-tocopherols (608 mg/kg seed) with efficient antioxidants. Therefore, sunflower is deemed as a plentiful source of vitamin E [10]. Compared with other oilseed meals, SFM is considered a good Ca, P, and vitamin B-complex [11]. Due to its low anti-nutritional and toxic compounds, sunflower proteins are considered an attractive alternative feed ingredient to replace SBM [12]. Researchers have extensively studied the potential functional properties of defatted oilseed meals [13]. Therefore, it is important to realize that the differences in its nutrient contents restrict the application of SFM in the poultry fed due to the different ways in the seeds' processing. SFM can be utilized in the diet of laying hens with no negative impact on the egg quality parameters [7,14–16]. By-products such as SFM contain high fiber and linoleic acid (a laying hen's fat source); the by-products are marketed for various world areas [17,18]. Fafiolu et al. [19] found that SFM is an excellent source of crude protein, ether extract, and amino acids, and it can be a substitute for SBM as feedstuff. SFM contains significant cell-wall components and high fiber content that may perform a crucial role in minimizing the blood cholesterol level. Baghban-Kanani et al. [8] revealed that the inclusion of SFM up to 20% of the laying hens' diets with multi-enzyme complex did not induce any negative impacts on the laying rate, egg quality traits, or antioxidant status. The partial substitution of SBM protein with SFM in Naked Neck hens' diets preserved successful efficiency and enhanced yolk color, showing that SFM was an economically viable substitute feed ingredient [20]. Earlier studies showed that the dietary SFM inclusion rates greater than 5% required lysine supplementation. SFM has a variable content of amino acids with lysine content that ranged from 0.56% to 0.66% and methionine content of 0.33% to 0.50% [14]. Lysine supplementation to the laying hens' diets containing SFM does not appear as crucial as in broilers' diets due to lower lysine requirement. Methionine, the first limiting amino acid, restricts egg weight, egg development, and egg mass [21–23].

Furthermore, SFM has potential environmental benefits in which the dietary inclusion of 20% SFM in the laying hens' diets significantly decreased ammonia and total nitrogen emissions [24]. The high fiber content of SFM is expected to have formed more fermented substrates in the gastrointestinal tract, leading to more significant microbial proteins [24]. Additionally, feeding poultry on SFM might have an indirect environmental impact by producing volatile fatty acids (VFA), which decreases the pH of the manure [25,26].

However, very few studies have assessed the dietary addition of SFM, as a supplier of polyunsaturated fatty acids (PUFA), in the laying hens' diets on the laying efficiency, yolk fatty acids (FA), and cholesterol concentration. Therefore, the current research is intended to assess the effect of dietary inclusion of SFM as a complementary protein resource on the laying performance, egg quality parameters, yolk fatty acids content, and nutrient digestibility of laying hens. The assumption examined was that the dietary inclusion of SFM might improve the production performance (egg production, egg mass, and feed conversion ratio), some egg quality characteristics (yolk height and yolk color), and enrich egg yolk with beneficial fatty acids (omega-3 fatty acids).

2. Materials and Methods

This study was permitted by the Local Experimental Animals Care Committee's Ethics Committee and done according to the rules of Kafrelsheikh University, Egypt. (No. 4/2016EC).

2.1. Chemical Composition of Sunflower Meal (SFM)

Sunflower meal was provided from the Egyptian raw material market in pellet form; this was ground before use. The chemical composition values used for soybean meal (SBM) and sunflower meal (SFM) were analyzed in the laboratory of feed analysis at Kafrelsheikh University, Egypt and the values recorded by national research council (NRC) [27] and shown in Table 1. The metabolizable energy content of SBM and SFM were calculated with the following equation [27]:

$$Men = 26.7 \times DM + 77 \times EE - 51.22 \times CF$$

where:

DM: dry matter, %.
EE: ether extract, %.
CF: crude fiber, %.

Table 1. Nutrient composition and metabolizable energy content of soybean meal and sunflower meal (% DM).

Nutrients	Soybean Meal	Sunflower Meal
DM, % [1]	92.06	91.20
Crude protein, %	46.0	36.00
ME, kcal/kg diet [2]	2350	1800
Calcium, %	0.3	0.40
Total Phosphorus, %	0.64	0.70
Ether extract, %	1.42	2.87
Crude fiber, %	5.6	17.00
Lysine, %	3.04	1.50
Methionine, %	0.66	0.91
Linolenic fatty acid, %	3.83	1.97

Analyzed values are mean of all replicates; [1] DM, dry matter; [2] ME, metabolizable energy.

2.2. Birds, Housing, and Experimental Design

A total of 162 Bovans laying hens, aged 60 weeks old (well beyond the laying peak and even the age at which most farms hens stop laying) with an average laying rate of 60.5%, was individually housed in cages in an open-sided structure under a 16-h light system, 8 h of darkness with LED light colors. A light intensity of 15 lux, however, controlled the dark period by closing the windows with blackout curtains. Laying hens (started lay at 20 weeks of age) were arbitrarily allotted into three dietary groups. Each group (54 laying hens) was randomly assigned into nine replicates; each replicate had six hens caged in Big Dutchman in regular dimensions of $40 \times 35 \times 60$ cm^3, in a double-sided battery cage. An automated nipple drinker was given for each cage. Birds were fed, on ad libitum basis, basal diet as the control, and two levels of SFM, 50 and 100 g/kg feed from 60 to 68 weeks of (thus from weeks 40 to 48 of laying). The composition of the experimental diets is presented in Table 2. Diets were calculated to meet the recommendations of NRC [27] for Brown Bovans laying hens.

Table 2. Ingredients and components of the experimental diets.

Ingredient	Diets, g/kg		
	Control	50 g SFM/kg	100 g SFM/kg
Yellow corn	635	650	605
Soybean meal, 46%	240	108	109
Corn gluten meal, 62%		60	37
Soybean oil	18	9	26
Di-calcium phosphate	20	19	17.8
Sunflower meal, 36%		50	100
Wheat bran		7	8
DL- methionine	2.1	1.6	1.6
L-lysine		3.2	3
Threonine	0.5	1.7	1.7
Limestone	72	73.6	74
NaCl	3	3	3
Premix *	4	4	4
Sodium bicarbonate	2.4	2.4	2.4
Potassium carbonate	3	6.5	6.5
Choline chloride		1	1
Total	1000	1000	1000
Calculated analysis **			
Crude protein, %	16.09	16.02	15.99
ME (kcal/kg diet)	2851	2850	2850
Calcium, %	3.26	3.29	3.29
Total phosphorus, %	0.71	0.70	0.71
Available phosphorus, %	0.46	0.46	0.47
Ether extract, %	4.46	3.68	5.12
Fiber, %	2.80	3.15	3.86
Lysine, %	0.88	0.89	0.89
Methionine, %	0.49	0.51	0.51
Chemical analysis			
Moisture, %	11.27	11.31	11.29
Crude protein, %	16.11	16.19	16.03
Ether extract, %	4.51	3.93	4.98
Fiber, %	2.93	3.22	3.97
Calcium, %	3.30	3.28	3.27
Total phosphorus, %	0.68	0.69	0.69

* Premix composition (units per kilogram of feed): vitamin A, 10,000 IU; vitamin D_3, 3,500 IU; vitamin E, 35 IU; menadione, 1.5 mg; vitamin B_1, 1.5 mg; vitamin B_2, 5 mg; vitamin B_5, 8 mg; vitamin B_6, 1.5 mg; vitamin B_{12}, 0.012 mg; folic acid, 0.5 mg; niacin, 30 mg; biotin, 0.06 mg; iodine, 0.8 mg; Cu, 10 mg; Fe, 80 mg; Se, 0.3 mg; Mn, 80 mg; Zn, 80 mg. ** Calculated according to NRC [27] for Brown Bovans laying hens. The diets were provided in mash form.

2.3. Performance Parameters

At the beginning (60 weeks of age) and the end (68 weeks of age) of the trial, the birds were weighed individually by ZIEIS Digital Bird Scale, A63SS-NMP, 0.05 Ounce Accuracy, 5000 Gram Capacity. Eggs were hoarded every day. The laying rate was calculated as hen-day (% hens-day) by applying the following equation (number of daily eggs produced per treatment/number of birds accessible in the treatment on that day × 100). Each egg weight was assessed and then utilized for all experimental times to evaluate the mean egg weight. The total egg mass was determined by laying rate by multiplying the weights of the eggs. As the hens were fed by an ad libitum system, the feed amount was added according to the catalog, and after seven days, the remaining feed was measured, and then the intake of feed was calculated on a cage base (a hen). Daily feed consumption per hen for all days during the trial was determined. The FCR (kg of feed/kg of eggs) was assessed utilizing egg production, egg weight, and feed intake.

2.4. Egg Quality Parameters

Egg quality parameters including egg length, egg width, egg shape, shell thickening, high albumin, high yolk, yolk width, yolk index, and yolk color score were undertaken and measured at the beginning of the experiment (60 weeks of age) and the end of the experiment (68 weeks of age). From each test, 30 eggs lay between 08:00 and 12:00 h were arbitrarily selected. A digital egg scale individually weighed eggs, accurate to 1/10th of a gram, 100 g maximum capacity, and the egg quality estimation was done on individual eggs, likewise the egg weight. The eggs were broken on the plate measurement stand egg Quality Microprocessor (EQM), and the albumen and yolk heights were determined. Yolk color score was measured utilizing the Roche yolk color fan method (DSM Yolk Color Fan, Basel, Switzerland). Eggshell thickness was performed by determining the thickness mean values taken at three locations on the egg (air cell, equator, and sharp end) utilizing a micrometer caliper (Mitutoyo, 0.01 to 20 mm, Tokyo, Japan).

2.5. Yolk Fatty Acid Content, Total Cholesterol, Vitamin E, and Ca Concentrations

At the beginning of the experiment (60 weeks) and the end (68 weeks), 30 eggs were collected per procedure to measure the content of FAs in the egg yolk, including myristic, palmitic, palmitoleic, stearic, oleic, veccenic, linoleic, linolenic, arachidonic, Eicosapentaenoic acid (EPA), Docosapentaenoic acid (DPA), Docosahexaenoic acid (DHA), yolk fat and total cholesterol. Analysis of the previous fatty acids was performed using a Shimadzu GC-4 CM (PFE) gas chromatograph fitted with a flame ionization detector (FID).

Before running the samples, a regular blend of methyl esters was examined under similar circumstances. The retention times of the unidentified methyl ester sample were compared with those of the standard. In the triangulation process, the quantity of methyl esters was assessed according to Radwan [28] and Saleh et al. [29].

Fatty acids were expressed as mg/100 g fat. For determination of vitamin E and Ca in the egg yolk, pooled samples were homogenized in a 0.054 mol/L dibasic sodium phosphate buffer amended to 7 pH with HCl. After being mixed with absolute ethanol and hexane, the upper layer α-T was evaporated and dissolved in ethanol before evaluation by HPLC3 (UV detector fixed at 290 nm). Egg yolk total cholesterol was measured through the extraction of fat from the egg yolk with chloroform and methanol admixture (2:1 vol:vol) methods according to Surai [30] and expressed as mg/100 g fat.

Calculation of the lipid quality indices including the atherogenic index (AI) and the thrombogenic index (TI) were performed following Ulbricht and Southgate [31]. The peroxidability index (PI) was assessed using the equation of Arakawa and Sagai [32].

2.6. Nutrient Digestibility

For digestibility tests, excreta were collected and weighted from each cage replicate over the last three days of the experiment. The feed intake and birds were weighted daily during these three days, and feces eliminated were collected, weighed, and placed in a freezer. Following the digestibility trial, all samples were dried in a drying oven at 60 °C for 24 h. Next, the whole dried samples were homogenized according to AOAC [33] and finely ground for examination. The crude protein substance in the diet and excreta was determined using the Kjeldahl method to determine the digestibility of nitrogen (CP, Method 968.06), the fat extract was calculated using the Soxhlet method (EE, Method920.39), crude fiber (CF, Method 932.09) and calcium (Ca, Method 985.35). The calculation was as follows:

Nitrogen digestibility (%) = (total nitrogen intake − total nitrogen excreted)/total nitrogen intake × 100.

2.7. Statistical Analysis

Statistically, the experimental results were analyzed using a one-way analysis of variance (ANOVA) (IBM SPSS Statistics Version 25.0. Armonk, NY, USA: IBM Corp). We contrasted the means of different treatments using Duncan's new multiple range test. The limit of significance was at $p < 0.05$.

3. Results

3.1. Laying Performance

Table 3 presents the impact of feeding SFM on the efficiency parameters of laying birds. Non-significant changes in final body weight, body weight gain, feed intake, egg weight, and egg mass were observed among the dietary groups. An increase in egg production ($p < 0.05$) was noted in laying hens fed the SFM diets compared with those fed the control diet. The percent of broken eggs was the lowest from laying hens-fed SFM ($p < 0.05$). Laying hens fed the SFM diets had better FCR ($p < 0.05$) when controlling the control one.

Table 3. Effect of feeding sunflower meal (SFM) on the production performance of laying hens.

Item	Experimental Diets			SEM	*p*-Value
	Control	50 g SFM/kg	100 g SFM/kg		
Initial body weight (60 wks.), g	1512.8	1516.1	1516.1	37.11	0.99
Final body weight, (68 wks.), g	1586.5	1587.2	1588.8	37.35	0.99
Body weight gain, g	73.8	71.1	72.6	1.42	0.75
Feed intake, g/day	116.6	116.5	116.4	0.39	0.98
Egg production, %	61.7 [b]	65.0 [a]	65.4 [a]	0.04	0.05
Egg weight, g	55.8	58.2	58.7	0.58	0.12
Egg mass, g of egg/hen/day	34.4	37.8	38.4	0.49	0.09
FCR, g feed/g egg	3.39 [b]	3.08 [a,b]	3.03 [a]	0.06	0.05
Broken egg, %	11.4 [a]	9.9 [b]	8.9 [b]	0.26	0.05

Values are presented as means ± SE of 60 per group. [a,b] Mean values with different superscripts in the same row are different at $p < 0.05$. FCR = feed conversion ratio.

3.2. Selected Egg Characteristics

The effect of feeding SFM on the selected egg characteristics at the beginning of the experiment (60 weeks of age) and the end of the experiment (68 weeks of age) is presented in Table 4. There was a non-substantial impact ($p < 0.05$) of the SFM levels on the egg quality characteristics, except for yolk color score and yolk height, which were higher ($p = 0.01$) for laying hens fed SFM concerning hens fed control.

Table 4. Impact of feeding SFM on the selected egg characteristics of laying hens.

Item	Experimental Diets			SEM	*p*-Value
	Control	50 g SFM/kg	100 g SFM/kg		
At week 40 of laying (60 wk of age)					
Egg length, cm	5.92	5.99	5.99	0.034	0.64
Egg width, cm	4.46	4.41	4.42	0.024	0.65
Egg shape index, %	75.33	73.62	73.79	0.005	0.29
Eggshell thickness, μm	327.9	328.6	328.9	4.06	0.19
Albumen height, cm	0.85	0.85	0.84	0.024	0.93
Yolk height, cm	2.03	2.03	2.04	0.016	0.89
Yolk width, cm	4.48	4.5	4.49	0.031	0.86
Yolk index, %	45.31	45.11	45.43	0.004	0.95
Yolk color score	6.7	6.7	6.7	0.13	0.99
At week 48 of laying (68 wk of age)					
Egg length, cm	5.85	5.93	6.41	1.17	0.36
Egg width, cm	4.44	4.47	4.46	0.028	0.89
Egg shape index, %	75.89	75.38	69.58	0.015	0.40
Eggshell thickness, μm	320.8	323.7	335.4	4.29	0.34

Table 4. *Cont.*

Item	Experimental Diets			SEM	*p*-Value
	Control	50 g SFM/kg	100 g SFM/kg		
Albumen height, cm	0.75	0.80	0.73	0.013	0.11
Yolk height, cm	1.83 [b]	2.13 [a]	2.22 [a]	0.056	0.01
Yolk width, cm	4.51	4.46	4.52	0.027	0.63
Yolk index, %	40.57	47.75	49.11	0.004	0.16
Yolk color score	6.6 [b]	7.8 [a]	7.9 [a]	0.145	0.01

Values are presented as means of 30 eggs per group and SEM for a total of 90 eggs from three study groups. [a,b] Mean values with distinct superscripts in the same row are different at $p < 0.05$.

3.3. Yolk Fat, Fatty Acid (FA) Content, Vitamin E, and Ca Contents in the Egg Yolk

Results concerning the effects of feeding SFM on egg yolk nutritional analysis in laying hens are shown in Table 5. The addition of SFM in the diets of laying hens did not influence the egg yolk fat content; however, it increased ($p < 0.05$) linoleic acid, linolenic acid, and oleic acid egg yolk content. On the other hand, palmitic acid's egg yolk concentration was decreased significantly by feeding SFM. Myristic, palmitoleic, stearic, vaccenic, arachidonic, eicosapentenoic, docosapentenoic, docosahexenoic acids AI, TI, and PI was not substantially affected by the dietary treatments. However, all fatty acids were not influenced at the beginning of the experiment. Interestingly, cholesterol level was significantly lowered by dietary treatments ($p < 0.05$).

Table 5. Effect of feeding SFM on the egg yolk fatty acid composition (%) of laying hens.

Item	Experimental Diets			SEM	*p*-Value
	Control	50 g FM/kg	100 g FM/kg		
At start (week 40 of laying)					
Myristic acid (C14:0)	0.23	0.22	0.23	0.02	0.63
Palmitic acid (C16:0)	24.5	24.52	24.45	0.62	0.75
Palmitoleic acid (C16:1)	2.78	2.77	2.93	0.23	0.62
Stearic acid (C18:0)	8.95	8.97	8.82	0.33	0.88
Oleic acid (C18:1 n-9c)	43.2	42.98	43.11	2.65	0.72
Vaccenic acid (C18:1 n-7)	1.95	1.92	1.94	0.21	0.81
Linoleic acid (C18:2 n-6)	14.44	14.62	14.51	0.92	0.58
Linolenic acid (ALA, C18:3 n-3)	0.52	0.53	0.56	0.032	0.41
Arachidonic acid (AA, C20:4 n-6)	1.81	1.91	1.81	0.091	0.42
Eicosapentenoic acid (EPA, C20:5 n-3)	0.088	nd	0.088	0.0001	0.82
Docosapentenoic acid (DPA, C22:5n-3)	0.111	0.111	0.112	0.001	0.64
Docosahexenoic acid (DHA, C22:6n-3)	0.867	0.866	0.869	0.002	0.34
AI	0.436	0.438	0.438	0.00611	0.98
TI,	0.932	0.931	0.931	0.00685	0.98
PI,	24.233	24.233	24.239	0.1147	1.00
After 8 weeks of the experiment (week 48 of laying)					
Myristic acid (C14:0)	0.22	0.23	0.24	0.021	0.57
Palmitic acid (C16:0)	26.8 [a]	23.11 [a,b]	20.07 [b]	1.42	0.042
Palmitoleic acid (C16:1)	2.94	2.58	2.32	0.36	0.37
Stearic acid (C18:0)	7.66	7.32	7.14	0.28	0.48
Oleic acid (C18:1 n-9c)	42.54 [b]	45.28 [a]	47.12 [a]	3.12	0.045
Vaccenic acid (C18:1 n-7)	1.94	1.93	1.93	0.23	0.79
Linoleic acid (C18:2 n-6)	14.34 [b]	15.51 [a,b]	17.45 [a]	1.01	0.048
Linolenic acid (ALA, C18:3 n-3)	0.42 [b]	0.52 [a]	0.56 [a]	0.052	0.042
Arachidonic acid (AA, C20:4 n-6)	1.8	1.8	1.79	0.092	0.92
Eicosapentenoic acid (EPA, C20:5 n-3)	0	0.027	0.028	0.00008	-
Docosapentenoic acid (DPA, C22:5n-3)	0.115	0.114	0.114	0.001	0.72
Docosahexenoic acid (DHA, C22:6n-3)	0.875	0.873	0.872	0.002	0.68

Table 5. *Cont.*

Item	Experimental Diets			SEM	*p*-Value
	Control	50 g FM/kg	100 g FM/kg		
AI,	0.427	0.422	0.421	0.00583	0.74
TI,	0.953	0.963	0.963	0.0145	0.86
PI,	24.495	24.501	24.783	0.2389	0.65
Yolk fat, g/100 g yolk	28.76	29.11	29.21	2.19	0.58
Total Cholesterol, mg/100 g yolk	137.07 [a]	130.60 [a,b]	122.47 [b]	2.45	0.04

Values are presented as means of 15 samples per treatment and SEM for 45 samples from all study groups; for "[a,b] Mean values with different superscripts in the same row are different at $p < 0.05$. nd = not detected.

The vitamin E and Ca contents in the egg yolk of laying hens fed the experimental diets were demonstrated in Table 6. Inclusion of SFM in the diets of laying hens increased ($p < 0.05$) vitamin E and Ca contents in the egg yolk.

Table 6. Effect of feeding SFM on vitamin E and calcium contents in egg yolk of laying hens at the end of the experiment.

Item	Experimental Diets			SEM	*p*-Value
	Control	50 g SFM/kg	100 g SFM/kg		
Vitamin E, mg/100 g	5.11 [b]	5.60 [a,b]	6.10 [a]	0.0116	0.01
Calcium content, mg/100 g	0.736 [b]	0.808 [a]	0.821 [a]	0.0058	0.03

Values are presented as means ± SE of 15 samples per group. [a,b] Mean values with different superscripts in the same row are different at $p < 0.05$.

3.4. Nutrient Digestibility

Table 7 reveals the effect of feeding SFM on nutrient digestibility in laying hens. The dietary treatments significantly increased the rate of CP ($p = 0.01$) and Ca ($p = 0.05$) digestibility. Interestingly, the addition of SFM in laying hens diets significantly decreased the digestibility of laying hens to EE ($p = 0.05$).

Table 7. Effect of feeding SFM on nutrient digestibility of laying hens at the end of the experiment.

Item	Experimental Diets			SEM	*p*-Value
	Control	50 g SFM/kg	100 g SFM/kg		
Crude protein, %	66.1 [b]	68.1 [a]	68.0 [a]	0.375	0.01
Crude fiber, %	33.3	36.3	35.2	0.633	0.15
Ether Extract, %	25.8 [a]	25.0 [a,b]	24.1 [b]	0.322	0.05
Calcium, %	32.3 [b]	36.0 [a,b]	37.1 [a]	0.911	0.05
Phosphorous, %	29.9	36.0	34.7	1.406	0.18

Values are presented as means of 15 samples per treatment and SEM for 45 samples from all study groups; for [a,b] Mean values with different superscripts in the same row are different at $p < 0.05$.

However, non-significant differences were detected in the CF and P digestibility between the dietary treatments.

4. Discussion

As the world's population grows, demand for eggs will continue to rise. To meet this demand sustainably will be a big challenge because of the traditional plant protein sources' high cost for layer hens diets. Furthermore, poultry nutritionists have been working for decades on sustainability in higher egg production. Using alternative plant protein sources like sunflower seed meal and others are innovative solutions for reducing the cost of the diets and improving the production, leading to the production and improvement of the environment.

4.1. Laying Performance

The current study's findings showed that feeding SFM significantly improved laying performance, broken egg ratio, and FCR for laying hens at a very late phase of laying (phase 2 of the production).

The available findings of the probable impacts of dietary inclusion of SFM on laying efficiency and FCR are questionable and contrasting. Several earlier studies have revealed that dietary inclusion of SFM had no adverse effect on live weight, feed intake, egg production, or FCR [2,7,8,15]. In contrast, other studies [4–6,14,34] showed that supplementation of SFM in the laying hens' diets improved the laying performance and FCR.

Additionally, Sunil [35] found a substantial increase in the rate of laying and FCR when SFM was incorporated in the diet at a concentration of 13% and attained maximum benefit. Due to the upsurge in the layer's body mass, body mass constancy in laying bird diets containing various protein resources can enhance laying performance [36].

Considering egg production percent and feed intake, FCR determination is possibly the largest single variable used in laying hens' economic assessments for the laying rate [37].

Additionally, the egg weight among experimental hens, was statistically similar. The average egg weight was also variable and compared favorably with laying hens' values recorded in the available literature [2,38]. For normal digestive function, a significant amount of fiber is needed. However, ingredients with high fiber content are limited in poultry diets due to their low energy content. The appropriate amount of crude dietary fiber in a realistic laying hen diet is between 35 and 45 g/kg [2]. Based on the dehulling degree, the rudimentary fiber of SFM seems to be the most critical component of poultry diets [39]. The enhancement in the laying performance of hens in the present trial might be ascribed to the use of high-protein and low-fiber SFM, and the added lysine contributes to the improved feed intake of laying hens. Seidavi et al. [9] indicated that SFM might be effectively included in the diets of laying hens up to 40% with an increase in egg production.

4.2. Egg Quality Parameters

In the present study, feeding SFM to laying hens did not influence the egg quality parameters. These results are inconsistent with Shi et al. [2], Baghban-Kanani et al. [8], Tsuzuki et al. [14], and Koçer et al. [16], who described non-substantial changes in the egg quality traits when laying birds were fed various dietary SFM levels. Meanwhile, dietary SFM supplementation substantially increased the yolk height and yolk color score. These results are close to Laudadio et al. [7], who observed that the egg yolk color record was improved when SFM with low fiber content was included in the layer diet concerning the SBM treatment layers. The effect of low-fiber SFM noted in our experiment on yolk color score may be linked to the number of natural pigments found in SFM. Previous studies [36,40] have shown an enhancement in yolk color as leguminous plant levels increased in the diets of laying hens.

On the other hand, adding dietary fat is essential as it accelerates the absorption of pigment and fat-soluble vitamins [41]. De Morais Oliveira et al. [20] indicated that the amount of lipids in the SFM diet augmented pigment absorption, resulting in improved yolk color. In contrast, Shi et al. [2] and Tsuzuki et al. [14] described no positive effect of dietary SFM on the egg yolk color.

4.3. Yolk Fat, Fatty Acids, Vitamin E, and Calcium Contents in Egg Yolk

The dietary addition of SFM, in the current study, increased the egg yolk contents of linoleic acid, α-linolenic acid, oleic acid, vitamin E, and Ca. Unsaturated FA plays a vital role in animal and human nutrition as they minimize metabolic problems such as cardiovascular diseases and diabetes [42]. It is commonly identified that sunflower is a good source of FA. In contrast, for high oleic sunflower oil, the reported contents for palmitic, stearic, linoleic, and oleic acids were 4.6%, 3.4%, 27.5%, and 62.8%, respectively. For ordinary sunflower oil, these values were 6.2%, 3.7%, 25.2%, and 63.1%, correspondingly [43]. Laudadio et al. [7] stated that substitution of SBM with SFM in layer diets did not cause any adverse impacts on egg production and egg quality, but modified the lipids contained in the yolk.

Comparably, Ebeid et al. [44] indicated that the increased α-linolenic acid in the eggs of laying hens might be achieved by introducing specific resources to the diets of laying hens like seed meals or oil sources. Additionally, sunflower seeds are exceptionally rich in α-tocopherols (608 mg/kg seed), which perform as potent antioxidants. Therefore, sunflower is believed to be a higher source of vitamin E. Nevertheless, heat inactivation of α-tocopherols is easier than p- and y-tocopherols, which are more common in soybean and cotton oil [45]. Furthermore, the protein obtained from SFM has a well-balanced composition of amino acids. SFM is considered a healthy source of Ca, P, and vitamin B-complex [46]. Our findings demonstrated that the inclusion of SFM in the diets of laying birds reduced the content of egg yolk cholesterol. Such results agreed with several previous studies [2,7,8], which recorded a substantial decrease in the egg yolk cholesterol when replacing SBM with SFM. This appears to appeal to consumers, as one of the primary health threat considerations associated with cardiac troubles is a higher circulating cholesterol level [47]. The hypo-cholesterolemic influence in serum and egg yolk of low-fiber SFM may be partially by diminishing the hepatic de novo lipogenesis.

Nevertheless, it is unidentified if SFM supplementation is efficient in decreasing the intestinal absorption of biliary cholesterol of laying hens, which regulates the whole-body cholesterol to reduce the cholesterol content in blood and egg yolk [47]. Additionally, a decrease in the yolk cholesterol content resulting from feeding low-fiber SFM may be partially due to the plant sterols present in sunflower with a hypo-cholesterolemic impact [48]. On the other hand, fiber's role in lowering cholesterol may be beneficial with the inclusion of SFM in the poultry diet. One possible mechanism in which SFM can perform its hypo-cholesterolemic effect is through bile acids. The cholic and deoxycholic bile acids are formed by hepatocytes from cholesterol and are conjugated with glycine and taurine correspondingly.

4.4. Nutrient Digestibility

In the present study, the substantial rise in the digestibility of CP and Ca in laying hens fed the SFM diets was an indication that these diets have met the birds' requirements and may have been caused by the reduction of anti-nutritional factors in the used SFM. Since sunflower has characteristics such as chlorogenic acid, which inhibits trypsin activity by 30%, the levels of chlorogenic acid of 40 g/kg in the sunflower seeds may have been enough to decrease the digestibility of the dietary protein. Consequently, the response to additional lysine, where about 43% of the chlorogenic acid was destroyed by heating at 100 °C or 135 °C for 5 h.

On the other hand, similar to soybean, cotton, and rapeseed meals, one advantage of SFM is that it does not include large levels of anti-nutritional factors [49,50]. Bedford and Classen [51] reported that the SFM content of raw fibers might be three times greater than SBM. The fibers' quantity, which originates from the cortex, is considered highly resistant to bacterial dilapidation in the gastrointestinal tract. This problem can be overcome by lowering the fiber content of SFM. Some promising findings have been recorded when meals are heat-treated [37], ground with pins [52], or air-classified [53,54]. Furthermore, laying hens have a more evolved digestive system than broilers in gut ability [15].

On the other hand, variations in complexity, chemical composition, treatment method, fusion levels, age of birds, and food preparation methods used in various studies may explain not always obtaining consistent results. Despite some contradictory findings, previous studies have observed that SFM is deemed a great supplier of protein in poultry diets to guarantee optimum poultry production [55]. Other considerations must also be considered including low fiber ratios, pelleting the feed, using oils, supplementing lysine, measuring protein solubility, and adding enzymes that suit the SFM NSP content to improve feed performance. Thus, further research regarding SFM quality factors that affect the digestibility of nutrients in laying hens should be investigated [56–59].

5. Conclusions

Increasing the dietary supplement level of SFM from 50 g/kg to 100 g/kg did not adversely impact body weight gain, feed intake, and egg mass. The dietary inclusion of SFM improved egg production, FCR, broken eggs rate, yolk color score, and yolk height of the laying hens. Dietary supplementation with SFM decreased egg yolk cholesterol, whereas vitamin E, Ca, linoleic acid, linolenic acid, and oleic acid contents in the egg yolk were increased. Furthermore, the addition of SFM in the diets of laying hens improved CP and Ca digestibility, but decreased the EE digestibility. Our results suggest that the inclusion of SFM, up to 100 tableg/kg, in the diets of laying bird at a late phase of laying could improve the production performance, selected egg characteristics, yolk linolenic acid, and oleic acid contents, and nutrient digestibility while decreasing egg yolk cholesterol.

Author Contributions: Conceptualization, A.A.S., A.E.-A., K.A., M.S., Y.Z.E. and M.H.A.; Methodology, A.A.S., A.E.-A., K.A., M.S., Y.Z.E. and M.H.A.; Software, S.S., M.M.S.; Validation, A.A.S., A.E.-A., K.A., Y.Z.E., M.H.A., S.S. and M.M.S.; Formal analysis, A.A.S., A.E.-A., K.A., Y.Z.E., M.H.A., S.S. and M.M.S.; Investigation, A.A.S., A.E.-A., K.A., Y.Z.E., M.H.A., S.S. and M.M.S.; Resources, A.A.S., A.E.-A., K.A., M.S., Y.Z.E. and M.H.A.; Data curation, S.S. and M.M.S.; Writing—original draft preparation, A.A.S., A.E.-A., K.A., Y.Z.E., M.H.A., M.S. and S.S.; Writing—review and editing, A.A.S., A.E.-A., K.A., Y.Z.E., M.H.A., S.S., M.S. and M.M.S. All authors have read and agreed to the published version of the manuscript.

Funding: The current work was funded by Taif University Researchers Supporting Project number TURSP-2020/09, Taif University, Taif, Saudi Arabia.

Institutional Review Board Statement: This study was declared by the Local Experimental Animals Care Committee's Ethics Committee and done according to the rules of Kafrelsheikh University, Egypt. (No. 4/2016EC).

Informed Consent Statement: Not applicable.

Data Availability Statement: All data sets collected and analyzed during the current study are available on request from the corresponding author.

Acknowledgments: We would like to thank Taif University Researchers Supporting Project number (TURSP—2020/09), Taif University, Taif, Saudi Arabia.

Conflicts of Interest: The authors declare no conflict of interest.

References

1. Alagawany, M.; Farag, M.; Abd-El-Hack, M.; Dhama, K. The practical application of sunflower meal in poultry nutrition. *Adv. Anim. Vet. Sci.* **2015**, *3*, 634–648. [CrossRef]
2. Shi, S.; Tong, H.; Zou, J. Effects of graded replacement of soybean meal by sunflower seed meal in laying hen diets on hen performance, egg quality, egg fatty acid composition, and cholesterol content. *J. Appl. Poult. Res.* **2010**, *21*, 367–374. [CrossRef]
3. Tüzün, A.; Olgun, O.; Yildiz, O. Effect of different dietary inclusion levels of sunflower meal and multi-enzyme supplementation on performance, meat yield, ileum histomorphology, and pancreatic enzyme activities in growing quails. *Animals* **2020**, *10*, 680. [CrossRef]
4. Laudadio, V.; Bastoni, E.; Introna, M. Production of low-fiber sunflower (*Helianthus annuus* L.) meal by micronization and air classification processes. *CyTA-J. Food* **2013**, *11*, 398–403. [CrossRef]
5. Laudadio, V.; Interona, M.; Lasstella, N.; Tufarelli, V. Feeding of low-fibre sunflower (*Helianthus annus* L.) meal as substitute of soybean meal in turkey rations: Effects on growth performance and meat quality. *J. Poult. Sci.* **2013**, *62*, 130–132. [CrossRef]
6. Vierira, S.; Penz, A.; Leboute, M.; Corteline, J. A nutritional evaluation of a high fiber sunflower meal. *J. Appl. Poult. Res.* **1992**, *1*, 382–388. [CrossRef]
7. Laudadio, V.; Ceci, E.; Lasstella, N.; Tufarelli, V. Effect of feeding low-fiber fraction of air-classified sunflower (*Helianthus annus* L.) meal on laying hen productive performance and egg yolk cholesterol. *Poult. Sci.* **2014**, *93*, 2864–2869. [CrossRef] [PubMed]
8. Baghban-Kanani, P.; Hossenintabar, B.; Azimi, Y.; Seidavi, T.; Laudadio, V.; Tufarelli, V. Effect of different levels of sunflower meal and multi-enzyme complex on performance, biochemical parameters and antioxidant status of laying hens. *S. Afr. J. Anim. Sci.* **2018**, *48*, 390–399. [CrossRef]
9. Seidavi, A.; Azizi, M.; Laudadio, V. Practical applications of agricultural wastes in poultry feeding in Mediterranean and Middle East regions. Part 2: Tomato, olive, date, sunflower wastes. *World's Poult. Sci. J.* **2018**, *74*, 443–452. [CrossRef]

10. Azizi, M.; Sedavi, A.; Ragni, M.; Laudadio, V.; Tufarelli, V. Practical applications of agricultural wastes in poultry feeding in Mediterranean and Middle East regions. Part 1: Citrus, grape, pomegranate and apple wastes. *World's Poult. Sci. J.* **2018**, *74*, 489–498. [CrossRef]

11. Garcia-Moreno, M.; Fernandez, M.; Velasco, L.; Perez-Vich, B. Genetic basis of unstable expression of high gamma-tocopherol content in sunflower seeds. *BMC Plant Biol.* **2012**, *12*, 71. [CrossRef] [PubMed]

12. González-Pérez, S.; Vereijken, M. Sunflower proteins: Overview of their physicochemical, structural and functional properties. *J. Sci. Food Agri.* **2007**, *87*, 2173–2191. [CrossRef]

13. Manamperi, W.; Wiwswnborn, P.; Chang, K. Effects of protein separation conditions on the functional and thermal properties of canola protein isolates. *J. Food Sci.* **2011**, *76*, 266–273. [CrossRef] [PubMed]

14. Tsuzuki, E.; Garcia, M.; Murankami, A. Utilization of sunflower seed in laying hen rations. *Braz. J. Poult. Sci.* **2003**, *5*, 179–182. [CrossRef]

15. Casartelli, E.; Filardi, R.; Junqueira, O. Sunflower meal in commercial layer diets formulated on total and digestible amino acids basis. *Braz. J. Poult. Sci.* **2006**, *8*, 167–171. [CrossRef]

16. Koçer, B.; Bozkurt, M.; Ege, G.; Tüzün, A.E. Effects of sunflower meal supplementation in the diet on productive performance, egg quality and gastrointestinal tract traits of laying hens. *Br. Poult. Sci.* **2020**. [CrossRef]

17. San, J.; Villamide, M. Nutritional evaluation of sunflower products for poultry as affected by the oil extraction process. *Poult. Sci.* **2001**, *80*, 431–437. [CrossRef]

18. Ditta, Y.; King, A. Recent advances in sunflower seed meal as an alternate source of protein in broilers. *World's Poult. Sci. J.* **2017**, *73*, 527–542. [CrossRef]

19. Fafiolu, A.; Oduguwa, O.; Jegede, A. Assessment of enzyme supplementation on growth performance and apparent nutrient digestibility in diets containing undecorticated sunflower seed meal in layer chicks. *Poult. Sci.* **2015**, *94*, 1917–1922. [CrossRef]

20. De Morais Oliveira, V.R.; Arruda, V.; Silva, J.; Souza, J. Sunflower meal as a nutritional and economically viable substitute for soybean meal in diets for free-range laying hens. *Anim. Feed Sci. Technol.* **2016**, *220*, 103–108. [CrossRef]

21. Elliot, M.A. Amino acid nutrition of commercial pullets and layers. In Proceedings of the Dairy Nutritional Strategies to Meet Economic and Environmental Challenges, Salt Lake City, UT, USA, 29–30 January 2008; p. 193. [CrossRef]

22. Saleh, A.A. Effects of fish oils on the production performances and polyunsaturated fatty acids and cholesterol levels of yolk in hens. *Emir. J. Food Agric.* **2013**, *25*, 605–612. [CrossRef]

23. Saleh, A.A. Effect of Low-Protein in Iso-Energetic Diets on Performance, Carcass Characteristics, Digestibilities and Plasma Lipids of Broiler Chickens. *Egypt. Poult. Sci. J.* **2016**, *36*, 251–262. [CrossRef]

24. Naseem, S.; King, A.J. Effects of multi-species lactobacillus and sunflower seed meal on nitrogen-containing compounds in laying hens' manure and biological components in blood serum. *J. Appl. Poult. Res.* **2020**, *29*, 130–141. [CrossRef]

25. Roberts, S.; Xin, H.; Kerr, J. Effects of dietary fiber and reduced crude protein on nitrogen balance and egg production in laying hens. *Poult. Sci.* **2007**, *86*, 1716–1725. [CrossRef]

26. Roberts, S.; Bregendahl, K.; Xin, H. Adding fiber to the diet of laying hens reduces ammonia emission. *Anim. Indus. Rep.* **2006**. [CrossRef]

27. National Research Council, NRC. *Nutrient Requirements of Poultry*; National Academies Press: Washington, DC, USA, 1994. [CrossRef]

28. Radwan, N.; Hassan, R.; Qota, E.; Fayek, H. Effect of natural antioxidant on oxidative stability of eggs and productive and reproductive performance of laying hens. *Int. J. Poult. Sci.* **2008**, *7*, 134–150. [CrossRef]

29. Saleh, A.; Ahmed, E.; Ebeid, A. The impact of phytoestrogen source supplementation on reproductive performance, plasma profile, yolk fatty acids and antioxidative status in aged laying hens. *Reprod. Domest. Anim.* **2019**, *54*, 846–854. [CrossRef] [PubMed]

30. Surai, P. Vitamin E and egg quality. *Qual. Eggs Egg Prod.* **1995**, *5*, 387–394. [CrossRef]

31. Ulbricht, T.L.V.; Southgate, D.A.T. Coronary heart disease: Seven dietary factors. *Lancet* **1991**, *338*, 985–992. [CrossRef]

32. Arakawa, K.; Sagai, M. Species differences in lipid peroxide levels in lung tissue and investigation of their determining factors. *Lipids* **1986**, *21*, 769–775. [CrossRef]

33. AOAC. *Official Method of Analysis*, 16th ed.; Association of Official Analytical Chemists: Washington, DC, USA, 1995. [CrossRef]

34. Furlan, A.; Mantovani, A.; Murakami, A. Use of sunflower meal in broiler chicks feeding. *Rev. Bras. Zootec.* **2001**, *30*, 158–164. [CrossRef]

35. Sunil, K. Usage of decorticated sunflower meal in feed formulation. *Int. J. Chem. Stud.* **2019**, *6*, 212–215. [CrossRef]

36. Laudadio, V.; Tufarelli, V. Treated fava bean (*Vicia faba* var. minor) as substitute for soybean meal in diet of early phase laying hens: Egg-laying performance and egg quality. *Poult. Sci.* **2010**, *89*, 2299–2303. [CrossRef] [PubMed]

37. Laudadio, V.; Tufarelli, V. Influence of substituting dietary soybean meal for dehulled-micronized lupin (*Lupinus albus* cv. Multitalia) on early phase laying hens production and egg quality. *Livest. Sci.* **2011**, *140*, 184–188. [CrossRef]

38. Al-Shami, M.A.; Salih, M.E.; Abbas, T.E. Effects of dietary inclusion of alfalfa (*Medicago sativa* L.) leaf meal and Xylan enzyme on laying hens' performance and egg quality. *Res. Opin. Anim. Vet. Sci.* **2011**, *2*, 14–18. [CrossRef]

39. Senkoylu, N.; Dale, N. Nutritional evaluation of a high-oil sunflower meal in broiler starter diets. *J. Appl. Poul. Res.* **2006**, *15*, 40–47. [CrossRef]

40. Güçlu, B.K.; Işcan, K.M.; Uyanik, F.; Eren, M.; Ağca, A.C. Effect of alfalfa meal in diets of laying quails on performance, egg quality and some serum parameters. *Arch. Anim. Nutr.* **2004**, *58*, 255–263. [CrossRef]

41. Costa, F.; Souza, J.; Goulart, R. Desempenho e qualidade dos ovos de poedeiras semipesadas alimentadas com dietas contendo óleos de soja e canola. *Rev. Bras. Zootec.* **2008**, *37*, 1412–1418. [CrossRef]

42. Simopoulos, A. Human requirement for N-3 polyunsaturated fatty acids. *Poult. Sci.* **2000**, *79*, 961–970. [CrossRef]

43. Ramos, M.J.; Fernández, C.M.; Casas, A.; Rodríguez, L.; Pérez, Á. Influence of fatty acid composition of raw materials on biodiesel properties. *Bioresour. Technol.* **2009**, *100*, 261–268. [CrossRef]

44. Ebeid, T.; Eid, Y.; Saleh, A.; Abd El-Hamid, H. Ovarian Follicular Development, Lipid Peroxidation, Antioxidative Status and Immune Response in Laying Hens Fed Fish Oil Supplemented Diets to Produce Omega-3 Enriched Eggs. *Animal* **2008**, *2*, 84–91. [CrossRef] [PubMed]

45. Dorrell, D.; Vick, B. Properties and processing of oilseed sunflower. *Sunflower Tech. Prod.* **1997**, *35*, 709–745. [CrossRef]

46. Morrison, W. Effects of refining and bleaching on oxidative stability of sunflower seed oil. *J. Am. Oil Chem. Soc.* **1975**, *52*, 513–522. [CrossRef]

47. Laudadio, V.; Ceci, E.; Lasstella, N.; Tufarelli, V. Dietary high-polyphenols extra-virgin olive oil is effective in reducing cholesterol content in eggs. *Lipids Health Dis.* **2015**, *14*, 5. [CrossRef] [PubMed]

48. Liu, X.; Zhao, H.; Thiessen, S.; House, J. Effect of plant sterol-enriched diets on plasma and egg yolk cholesterol concentrations and cholesterol metabolism in laying hens. *Poult. Sci.* **2010**, *89*, 270–275. [CrossRef]

49. Karunajeewa, H.; Abu-Serewa, S.; Tham, H.; Eason, P. The effects of dietary level of sunflower seeds and lysine on egg quality and laying performance of White Leghorn hens. *J. Sci. Food Agric.* **1987**, *41*, 325–333. [CrossRef]

50. Senkoylu, N.; Dale, N. Sunflower meal in poultry diets: A review. *World's Poult. Sci. J.* **1999**, *55*, 153–174. [CrossRef]

51. Bedford, M.; Classen, H. Reduction of intestinal viscosity through manipulation of dietary rye and pentosanase concentration is affected through changes in the carbohydrate composition of the intestinal aqueous phase and results in improved growth rate and food conversion efficiency of broiler chicks. *J. Nutr.* **1992**, *122*, 560–569. [CrossRef]

52. Wu, Y.; Nichols, N. Fine grinding and air classification of field pea. *Cereal Chemist.* **2005**, *82*, 341–344. [CrossRef]

53. Saleh, A.A.; Elnagar, A.M.; Eid, Y.Z.; Ebeid, T.A.; Amber, K.A. Effect of feeding wheat middlings and calcium lignosulfonate as pellet binders on pellet quality growth performance and lipid peroxidation in broiler chickens. *Vet. Med. Sci.* **2020**, *7*, 194–203. [CrossRef]

54. Saleh, A.A.; Zaki, A.; El-Awady, A.; Amber, K.; Badwi, N.; Eid, Y.; Ebeid, T.A. Effect of substituting wheat bran with cumin seed meal on laying performance, egg quality characteristics and fatty acid profile in laying hens. *Vet. Arhiv* **2020**, *90*, 47–56. [CrossRef]

55. Wu, Y.; Abbott, P. Protein enrichment of defatted salicornia meal by air classification. *J. Am. Oil Chem. Soc.* **2003**, *80*, 167–169. [CrossRef]

56. Gunawardena, C.; Zijlstra, R.; Beltranena, E. Characterization of the nutritional value of air-classified protein and starch fractions of field pea and zero-tannin faba bean in grower pigs. *J. Anim. Sci.* **2010**, *88*, 660–670. [CrossRef]

57. Saleh, A.A.; Kirrella, A.A.; Dawood, M.; Ebeid, T.A. Effect of dietary inclusion of cumin seed oil on the performance, egg quality, immune response and ovarian development in laying hens under high ambient temperature. *J. Anim. Physiol. Anim. Nutr.* **2019**, *103*, 1810–1817. [CrossRef] [PubMed]

58. Saleh, A.A.; Hayashi, K.; Ijiri, D.; Ohtsuka, A. The influence of dietary supplementation with Aspergillus awamori and feeding canola seed on the growth performance and meat quality in male broilers chickens. *J. Anim. Sci.* **2015**, *86*, 305–311. [CrossRef] [PubMed]

59. Saleh, A.A.; Paray, P.A.; Dawood, M.O. Olive Cake Meal and Bacillus licheniformis Impacted the Growth Performance, Muscle Fatty Acid Content, and Health Status of Broiler Chickens. *Animals* **2020**, *10*, 695. [CrossRef]

 *sustainability*

Article

Slow-Release Urea as a Sustainable Alternative to Soybean Meal in Ruminant Nutrition

Saheed A. Salami [1], Maria Devant [2], Juha Apajalahti [3], Vaughn Holder [4], Sini Salomaa [3], Jason D. Keegan [5] and Colm A. Moran [6,*]

1 Solutions Deployment Team, Alltech (UK) Ltd., Stamford PE9 1TZ, UK; saheed.salami@alltech.com
2 Ruminant Research Department, IRTA Torre Marimon, Caldes Montbui, 08140 Barcelona, Spain; maria.devant@irta.cat
3 Alimetrics Research Ltd., 02920 Espoo, Finland; j.apajalahti@alimetrics.com (J.A.); sini.t.salomaa@gmail.com (S.S.)
4 Centre for Applied and Animal Nutrition, Alltech Inc., Nicholasville, KY 40356, USA; vholder@alltech.com
5 Regulatory Affairs Department, Alltech European Bioscience Centre, A86 X006 Meath, Ireland; jasondjdk@gmail.com
6 Regulatory Affairs Department, Alltech SARL, 14500 Vire, France
* Correspondence: cmoran@alltech.com; Tel.: +33-687-508-761

Citation: Salami, S.A.; Devant, M.; Apajalahti, J.; Holder, V.; Salomaa, S.; Keegan, J.D.; Moran, C.A. Slow-Release Urea as a Sustainable Alternative to Soybean Meal in Ruminant Nutrition. *Sustainability* **2021**, *13*, 2464. https://doi.org/10.3390/su13052464

Academic Editors: Nikola Puvača, Vincenzo Tufarelli and Eva Voslarova

Received: 11 January 2021
Accepted: 20 February 2021
Published: 25 February 2021

Publisher's Note: MDPI stays neutral with regard to jurisdictional claims in published maps and institutional affiliations.

Abstract: Three experiments were conducted to evaluate the feasibility of using a commercial slow-release urea product (SRU; Optigen®, Alltech Inc., Nicholasville, KY, USA) as a partial replacement for vegetable protein sources in cattle diets. The first experiment was an in vitro rumen fermentation that evaluated the effect of replacing soybean meal (SBM) nitrogen with nitrogen from either SRU or free urea in diets varying in forage:concentrate ratios. The second experiment examined the effect of replacing SBM with SRU on in situ dry matter and nitrogen degradability in the rumen. In the third experiment, a feeding trial was conducted to evaluate the effect of replacing SBM (0% as-fed SRU) with 1% or 3% as-fed SRU on feed carbon footprint (CFP; total greenhouse gas emissions associated with the life cycle of feed raw materials) and the toxicity potential of SRU in growing beef cattle. Results showed that replacing SBM with SRU up to 1.3% did not negatively affect in vitro rumen fermentation parameters. Supplementing SRU favourably decreased ruminal accumulation of ammonia and lactic acid when compared to free urea. There was no significant effect on effective rumen degradability of dry matter and nitrogen when one-third of SBM was replaced by SRU in the in situ study. Compared with the 0% SRU diet, feed CFP decreased by 18% and 54% in 1% SRU and 3% SRU diets, respectively. Additionally, feeding up to 3% SRU diet to beef cattle did not affect health and intake, and blood hematological and biochemical indices were within the physiological range for healthy bulls, suggesting no indication of ammonia toxicity. Overall, these results indicate that SRU can be used as a sustainable alternative to partially replace vegetable protein sources in ruminant diets without compromising rumen function and health of ruminants.

Keywords: cattle; vegetable protein; urea; rumen fermentation; sustainability

1. Introduction

Ruminants play a crucial role in converting human-inedible resources to high-quality protein (meat and milk) which is vital for meeting the nutritional needs of humans [1]. Increasing population growth, household income and urbanization have contributed significantly to a growing demand for animal protein, which has been projected to increase by 70% between 2005 and 2050 [2,3]. In recent times, the sustainability of the ruminant sector has attracted intense public concerns due to the environmental impacts of the sector. Thus, the ruminant sector is faced with the serious challenge of meeting the growing demand for animal protein while reducing its environmental impacts. The environmental impacts of the ruminant sector are largely due to manure nitrogen excretion and greenhouse gas

(GHG) emissions originating mainly from enteric fermentation, as well as feed production, processing and transport [3]. The low nitrogen utilization efficiency of ruminants increases manure nitrogen excretion, which can negatively affect air quality through ammonia volatilization, water quality through nitrate leaching and eutrophication and global warming through the release of nitrous oxide [4,5]. Opio, et al. [6] estimated that GHG emission from the ruminant sector was approximately 5.7 gigatonnes CO_2-eq, of which beef and dairy cattle production contribute about 4.6 gigatonnes CO_2-eq, representing 65% of the anthropogenic emissions from the livestock sector. Feed emission is the second-largest source of ruminant emissions, contributing about 36% of emissions associated with beef and milk production [3]. Thus, reducing the carbon footprint (CFP) of ruminant feeds is crucial for improving the resource efficiency and environmental performance of the ruminant sector [5,7,8].

Ruminants have the unique capability of converting dietary non-protein nitrogen (NPN) sources, such as urea, into high-quality protein via microbial protein synthesis in the rumen. Traditionally, the low cost of urea has increased its use as a partial substitute for vegetable protein sources, such as SBM (soybean meal) and rapeseed meal, to supply rumen-degradable protein (RDP) in ruminant diets. Moreover, the use of urea products could reduce the effect of land-use changes and GHG emissions associated with the cultivation and processing of plant protein sources, thus improving the resource efficiency and carbon footprint of ruminant production [9,10]. However, the utilization of feed-grade (unprotected) urea in ruminant nutrition is limited due to its rapid ruminal hydrolysis to ammonia, often exceeding the rate of carbohydrate fermentation in the rumen [11]. Asynchronization between ammonia production and fermentable energy availability in the rumen negatively affects the efficiency of microbial protein synthesis. This condition decreases the amount of rumen microbial protein outflow which may reduce the availability of metabolizable protein required to meet the maintenance and production needs of ruminants [12]. The rapid ruminal hydrolysis of feed-grade urea could also reduce the nitrogen utilization efficiency of ruminants, resulting in negative environmental impacts associated with increased nitrogen excretion and ammonia emission from manure [12]. Furthermore, the rapid degradation of dietary urea in the rumen can increase the blood ammonia concentration and increase the risk of ammonia toxicity and related negative health impacts in ruminants [13]. As a consequence, dietary urea supplementation is typically limited to 1% of total dietary dry matter or 0.3 g/kg body weight (BW)/day in animals adapted to urea consumption and fed a diet with an adequate supply of readily fermentable carbohydrates [14].

To alleviate the limitations associated with the use of feed-grade urea, several research efforts have been dedicated to developing coated urea products with a reduced rate of hydrolysis in the rumen and therefore improving the efficiency of microbial protein synthesis, while minimizing nitrogen excretion and toxic effects in ruminants [11]. In earlier research, Owens, et al. [15] estimated that 900 g of urea consumption in tung- and linseed-oil-coated urea would be required to cause ammonia toxicity in steers. Similarly, other slow-release forms of urea supplemented at 1.2–1.9% of the diet have been shown to exert no detrimental effects on the health or performance of ruminants with improved nitrogen use efficiency [16–18]. Another commercial slow-release urea product (SRU; Optigen®, Alltech Inc., Nicholasville, KY, USA) involves the use of urea prills embedded in a lipid matrix. The SRU has demonstrated a reduction in the rate of urea degradation in the rumen [18,19] and led to a more consistent concentration of ruminal ammonia compared to feed-grade urea [20]. Indeed, the rate of microbial urea hydrolysis must proceed in synchrony with bacterial ammonia assimilation and fermentable energy availability to maximize microbial protein synthesis in the rumen.

Data from the Dutch FeedPrint software developed by Wageningen University and Research, Wageningen, The Netherlands [21] indicate that the isonitrogenous replacement of soybean meal nitrogen with SRU will reduce the carbon footprint of the diet. However, feeding strategies aimed at utilizing SRU to replace vegetable protein sources must be achieved without impairing rumen function and production performance, and without significant risk of toxicity to animals. In this paper, three studies were conducted to evaluate the feasibility of using SRU to replace vegetable protein sources in cattle diets. The first experiment examined the effect of dietary SRU on in vitro rumen fermentation parameters at different forage:concentrate ratios to understand its potential application across beef diets high in forage or grain. The second experiment was designed to evaluate the effect of replacing vegetable protein with SRU on dry matter and nitrogen degradability using in situ techniques. The third experiment evaluated the impacts of replacing SBM with dietary SRU on feed CFP and the toxicity potential of SRU when fed up to 3% of cattle diet. To test the toxicity potential of SRU, the maximal supplemented SRU dose (i.e., 3%) selected in this experiment was three-fold higher than the maximum dose recommended for free urea [14], but lower than the expected threshold (i.e., 4.8%) for toxicity [15].

2. Materials and Methods

2.1. Experiment 1: In Vitro Evaluation to Determine the Effect of Slow-Release Urea on Rumen Fermentation

2.1.1. Diets, Treatments, and Doses

In this experiment, an in vitro rumen fermentation model developed by Alimetrics Research Ltd. (Espoo, Finland) [22,23] was used to compare the effect of replacing SBM nitrogen with nitrogen from either SRU or urea on the ruminal fermentation of high-forage and high-grain rations. Two basic diets were used in the rumen fermentation study: a high-forage ration consisting of 200 mg DM (dry matter) of wheat and 800 mg DM of grass silage and a high-grain ration consisting of 600 mg DM of wheat and 400 mg DM of grass silage. The total amount of feed introduced in each fermentation vessel was 1000 mg. Each nitrogen supplement, SBM, SRU (Optigen®, Alltech Inc., Nicholasville, KY, USA) and urea, were introduced at three doses. Notably, the SRU (Optigen®) consists of urea evenly coated with a semi-permeable vegetable fat matrix containing 88% urea (41% N, 256% crude protein (CP)) and 11–12% fat [24]. The fat coating in the SRU slows the dissolution of urea, reducing the rate of urea conversion to ammonia in the rumen [19].

The starting point for dose determination was that 20, 60 or 100 mg of each basic ration was replaced by the corresponding amount of SBM (2, 6 and 12% of ration DM). The total nitrogen analysis of the SBM preparation revealed that the amount of nitrogen thus added was 1.07, 3.22 and 5.35 mg, respectively. The doses of SRU and urea were calculated so that the amount of nitrogen added was the same as in 20, 60 and 100 mg of SBM. Since the nitrogen densities in SRU and urea were significantly higher than in SBM, smaller proportions of basic diets were replaced by these nitrogen sources. For SRU, the doses were 2.6, 7.8 and 13.1 mg (0.26, 0.78 and 1.31% of ration DM) and for urea, the doses were 2.3, 6.9 and 11.5 mg in the 1000 mg amount of diet (0.23, 0.69 and 1.15% of ration DM). Each of the 20 treatments was fermented in 5 replicate vessels. Thus, the total number of fermentation vessels was 100. Table 1 summarizes the treatment combinations and dosages used in this study.

2.1.2. Rumen Fermentation Simulation

Individual feed components and test products were weighed in 120 mL serum bottles, flushed with CO_2 that had passed through a hot copper catalyst for O_2 scavenging and sealed with thick butyl rubber stoppers. In each simulation vessel, 38 mL of anaerobic, reduced, temperature-adjusted (38 °C) buffer solution was added under oxygen-free CO_2 flow. Ruminal fluid was obtained from a rumen-fistulated early-lactation Ayshire cow fed *ad libitum* grass silage (13 kg dry matter with energy 10.8 MJ/kg and CP 16%) and a compound feed at 8 kg DM/day (Opti-Maituri 26, Lantmännen Feed Oy, Turku, Finland; energy 12.8 MJ/kg, CP 26%). Finally, 2 mL of fresh, strained rumen fluid was added to the serum bottles, where the final volume was 40 mL. This inoculation started the actual fermentation. Inoculation time for each vessel was registered and considered when sampling and stopping the fermentation. Details of the protocol are described by Meissner, et al. [23] and Apajalahti, et al. [22].

Table 1. Treatments used for the evaluation of slow-release urea in in vitro rumen fermentation.

Diet Number	Ration [1]	Nitrogen Supplement	Dose, mg (Nitrogen Levels) [2]	Replicate Fermentations
1		None	-	5
2			20 (low)	5
3		SBM	60 (medium)	5
4			100 (high)	5
5	High-forage		2.6 (low)	5
6		SRU	7.8 (medium)	5
7			13.1 (high)	5
8			2.3 (low)	5
9		Urea	6.9 (medium)	5
10			11.5 (high)	5
11		None	-	5
12			20 (low)	5
13		SBM	60 (medium)	5
14			100 (high)	5
15	High-grain		2.6 (low)	5
16		SRU	7.8 (medium)	5
17			13.1 (high)	5
18			2.3 (low)	5
19		Urea	6.9 (medium)	5
20			11.5 (high)	5
			Total	100

SBM: soybean meal, SRU: slow-release urea. [1] High-forage ration: 20:80, wheat: grass silage ration; high-grain ration: 60:40, wheat: grass silage ration. [2] Low, medium and high nitrogen levels correspond to 1.07 mg N, 3.22 mg N and 5.35 mg N, respectively, in the total 1 g of diet.

2.1.3. Analytical Procedures

Rumen fermentation simulation was continued for 24 h at 38 °C. During fermentation, the total gas production was measured after 3, 6, 9, 12 and 24 h of simulation fermentation to reveal the general metabolic activity of rumen microbes. Volatile fatty acids (VFA), lactic acid and pH were measured in all simulation vessels at 0, 9 and 12 h by taking 1 mL of the inoculum from each vessel through the rubber stopper using a 1 mL syringe and needle. The short-chain fatty acids (lactic acid and VFA) were analyzed by gas chromatography using a packed column for the analysis of free acids as described in detail by Apajalahti, et al. [22]. In brief, the following volatile fatty acids were analyzed: acetic, propionic, butyric, valeric, isobutyric, 2-methylbutyric and isovaleric acid. The main non-volatile fatty acid quantified was lactic acid. In this paper, all these acids are cumulatively referred to as short-chain fatty acids (SCFA). In the gas chromatographic analysis, a glass column packed with 80/120 Carbopack B-DA/4% Carbowax stationary phase was used. The use of this column enabled quantification of the free acids, with no derivatization. Pivalic acid was used as an internal standard (Sigma-Aldrich, St. Louis, MO, USA). The analytical chromatography run was isothermal at 175 °C (Agilent Technologies, Santa Clara, CA, USA), helium was used as a carrier gas and analytes were detected by flame ionization.

Total bacteria density was analyzed at 12 h in all the simulation vessels. Bacterial samples were fixed with formaldehyde, stained with DNA-specific dye SYTO® 24 and the total bacteria were enumerated by flow cytometry using settings adjusted to the counting of bacteria. The details of the method have been described by Apajalahti, et al. [25]. The samples from simulation vessels were analyzed for ammonia at 0, 3, 9 and 12 h. Ammonia analysis was performed by using a colorimetric method described by Weatherburn [26] modified from the Berthelot reaction [27]. The method was based on the reaction of phenol and hypochlorite with ammonia, resulting in color formation and the color intensity was measured with a spectrophotometer.

2.2. Experiment 2: In Situ Evaluation to Determine the Effect of Slow-Release Urea on Feed Degradability in the Rumen

2.2.1. Diets, Treatments and In Situ Incubation

Two rumen-fistulated cows in mid-lactation were fed with grass silage (CP = 16.4% of DM) and a compound feed mixture (CP = 12.4% of DM) of hay, oats, wheat and protein-rich meals (soy, pea and/or rapeseed), and were used to examine the effect of SRU (i.e., Optigen®) on in situ dry matter (DM) and nitrogen degradation. Two test diets (control and SRU) were formulated as shown in Table 2. The control diet had no SRU whereas the SRU diet contained one-third of the SBM in the control diet, replaced with an isonitrogenous amount of SRU (0.75%). Thus, a cow consuming 20 kg DM/day of a total mixed ration would have been provided 150 g of SRU. The CP content of both diets was 19.9%. Dacron bags (10 cm × 20 cm; 53 μm pore size; ANKOM Technology Corp., Fairport, NY, USA) containing 5 g of the corresponding test diets were incubated in the rumen of the two cows. At 0 h, there were 6 replicate bags and 3 replicates for the other time points. Bags were withdrawn from the rumen after 0, 2, 5, 8, 16, 24 and 48 h of incubation. The bags were immediately dipped in cold water to stop fermentation, after which the bags were rinsed with water and the excess water was removed with a spin dryer and the bags were freeze-dried.

Table 2. Composition of test feeds used for in situ rumen degradability.

Diet Component	Nitrogen Content (g/kg DM)	Crude Protein (g/kg DM)	DM (%)	% of Diet DM	
				Control	SRU
Soybean meal	82.3	514	88.9	15	10
SRU	410	2563	100	0	0.75
Compound feed [1]	19.9	124	90	44	45.25
Grass haylage	26.2	164	72.1	41	44

SRU: slow-release urea. [1] Compound feed (CP—12.4% of DM) containing hay, oats, wheat and protein-rich meals.

2.2.2. Analysis of Dry Matter and Nitrogen Degradation

Residual dry matter in the Dacron bags was determined by weighing the bags after drying and dry matter (DM) residues in three replicate bags were determined. The nitrogen content in the feed residues was determined by the Kjeldahl method [28]. In situ degradation curves of DM and nitrogen were fitted to a non-linear model [29] using the equation:

$$Y = a + b \left[1 - e^{-K_d(t)} \right]$$

where Y = ruminal degradation of DM or nitrogen (%); a = rapidly soluble fraction that disappeared at 0 h after the rinsing procedure; b = potentially degradable fraction; K_d = constant rate of degradation of fraction b; and t = time of incubation (h). The un-degradable fraction, c, was calculated as $100 - (a + b)$. Effective rumen degradability (ERD) was determined by the equation:

$$ERD = a + \left[bK_d / \left(K_d + K_p \right) \right]$$

where a, b and K_d are the degradation constants described previously [29] and K_p is the passage rate from the rumen (%/h) assumed fixed at 0.06/h. Degradation constants of DM and nitrogen described above were estimated using the NLIN procedure of SAS (Statistical Analysis Systems, Cary, NC, USA) [30].

2.3. Experiment 3: In Vivo Trial to Evaluate Feed Carbon Footprint and Toxicity Potential of Slow-Release Urea in Cattle

The research protocol and animal care followed the European Union Directive 2010/63/EU on the protection of animals used for experimental or other scientific purposes and were managed according to the regulations of the Animal Care Committee of the Institute of Agrifood Research and Technology (IRTA), Spain. The animal experiment was conducted at the Corporació Alimentaria de Guissona Experimental Station, Lleida, Spain.

2.3.1. Animals, Diets and Experimental Design

Twenty-four growing Holstein bulls (3–4 months of age) were commercially-sourced and subjected to the following preventive treatments before the start of the experiment: Draxxin® (intravenous injection of tulathromycin, 2.5 mg/kg BW, Pfizer Animal Health, Parsippany-Troy Hills, NJ, USA) to prevent intercurrent respiratory disease; Vectimax® (subcutaneous injection of ivermectin, 0.2 mg/kg BW, Esteve Veterinaria, Barcelona, Spain) for internal and external parasites; CattleMaster® (subcutaneous application of a vaccine against IBR, BVD, PI3, BRSV, 2 mL/animal, Pfizer Animal Health), to prevent relevant viral respiratory diseases affecting young ruminants; Vecoxan® (oral Diclazuril, 1 mg/kg BW, Esteve Veterinaria) to prevent coccidiosis.

The growing Holstein bulls were blocked by liveweight (128.1 ± 14.2 kg) and randomly assigned to one of three treatments: a control diet, basal mash feed (CON, 0% SRU) or the basal diet reformulated to include SRU (i.e., Optigen®) at a rate of 1% (1% SRU) or 3% (3% SRU) of the diet on an as-fed basis. Energy and CP levels were maintained in the SRU diets by replacing SBM with SRU, barley grain meal and corn grain meal; the ingredient and nutrient composition of diets are presented in Table 3. The SRU was added in meal form to the complete feed and thoroughly mixed. Diets contained no antibiotics or other growth promoters. Trace minerals and other nutrients were supplied at nutritional concentrations according to the NRC requirements [31]. Basal diets were calculated to be iso-nutritive, to meet the nutrient requirements recommended for growing ruminants [31]. Bulls were allocated to an individual, partially slatted pens (2.40 × 1.45 m), with two feeders in each pen (one for mash feed and one for straw), which were hand-filled daily. Animals were allowed *ad libitum* access to both the mash feed and straw and an automatic watering device was available in each pen. The experimental feeding period lasted for 42 days. All animal handling and laboratory staff were blinded to the study diets.

Table 3. Ingredient composition of the experimental mash diets of growing bulls supplemented with slow-release urea (SRU) in replacement for soybean meal.

Ingredient (%)	0% SRU	1% SRU	3% SRU
Corn grain meal	38.19	40.96	48.00
Wheat middlings	3.02	3.02	2.00
Barley grain meal	17.02	19.02	22.61
Wheat	2.51	2.51	2.00
Beet pulp	7.99	7.99	9.99
Dried alfalfa	8.33	8.33	6.52
Palm oil	2.46	2.85	3.27
Soybean meal	18.38	12.16	0.00
Calcium carbonate	0.80	0.86	0.80
Sodium bicarbonate	0.60	0.60	0.60
Salt	0.40	0.40	0.40
Bicalcium phosphate	0.00	0.00	0.51
Vitamin/mineral premix 809 [1]	0.30	0.30	0.30
Slow Release Urea	0.00	1.00	3.00
Analyzed nutrient content (% DM)			
DM	87.46	87.81	87.51
Crude protein	18.22	18.03	18.28
Ether extract	4.28	5.39	6.15
Neutral detergent fibre	26.17	25.56	26.14
Ash	5.99	5.62	4.86
Urea	-	<0.1	0.88
Calculated NSC	53.2	53.1	53.3

DM: dry matter; NSC: non-structural carbohydrates. [1] Vitamin/mineral premixture: Calcium 0.0102%; Phosphorus 0.0612%; Magnesium 29.02%; Sodium 0.0139%; Selenium 150 mg/kg; Cobalt 240 mg/kg; Iodine 256 mg/kg; Manganese 15,525 mg/kg; Zinc 20,350 mg/kg; Copper 2500 mg/kg Vitamin A (retinyl acetate) 5,000,000 IU/kg; Vitamin D3 (cholecalciferol) 1,000,000 IU/kg; Vitamin E (di-alpha tocopheryl acetate) 50,000 mg/kg.

Animal liveweights were determined on d 0, 21 and 42. Individual intake of the experimental mash diets was recorded daily, while the straw intake was recorded weekly. Animal health, culls and mortality were recorded daily. Blood samples were collected from each animal on d 0 and 42 (1 h post-feeding) for routine hematological and biochemical analysis (total blood cell counts, hemoglobin, hematocrit, aspartate aminotransferase, alanine aminotransferase, gamma-glutamyl transferase, glucose, uric acid, albumin, total protein, urea, and ammonia). Blood samples were taken via jugular venipuncture using a vacutainer and an 18 G needle. For hematology, 4 mL of blood was collected in EDTA vacutainer tubes (BD, Franklin Lakes, NJ, USA), inverted and stored at 5 °C until analysis. For biochemistry analysis, 10 mL of blood was collected in spray-dried clot activator vacutainer tubes (BD, Franklin Lakes, NJ, USA). For glucose analysis, 4 mL of blood was collected in sodium fluoride and potassium oxalate vacutainer tubes (BD, Franklin Lakes, NJ, USA). For ammonia analysis, 4 mL of blood was collected in EDTA dipotassium salt vacutainer tubes (BD, Franklin Lakes, NJ, USA). The vacutainer tubes for biochemistry, glucose and ammonia were then centrifuged at $1500 \times g$ at 4 °C for 15 min and the serum from each tube was equally divided between three Eppendorf tubes.

2.3.2. Chemical Analysis

Proximal analysis and the urea concentration of each experimental diets was determined using the following methods: Crude protein (AOAC 988.05); Crude fat (AOAC 920.39); Ash (AOAC 642.05); Moisture (AOAC 925.04); Neutral Detergent Fiber [32]; Urea (AOAC 967.07). Plasma glucose concentration was determined following the hexokinase method, serum aspartate aminotransferase and alanine aminotransferase following recommended IFCC reference method, without pyridoxal phosphate addition, serum urea following GLDH method, serum uric acid following uricase/peroxidase method, serum ammonia method by molecular absorption spectrometry, serum albumin following bromocresol green method, and serum total protein following biuret method (OSR, Winston-Salem, NC, USA). For hematology analyses, the ADVIA 120 Hematology System developed by the manufacturer (Siemens Healthcare GmbH, Erlangen Germany) was used. The following parameters were measured using flow cytometry methods and specific staining such as peroxidase and basophilic staining: red cell count, hemoglobin, packed cell volume, mean corpuscular volume, mean corpuscular hemoglobin, cell hemoglobin concentration mean, white cell count, percentage and number of neutrophils, lymphocytes, monocytes, eosinophils and platelet count.

2.3.3. Calculations

The impact of replacing SBM with SRU on feed CFP was determined by calculating the sum of the CFP of feed raw materials in 0% SRU, 1% SRU and 3% SRU diets. The CFP values (including land-use changes) of feed raw materials were retrieved from the Dutch FeedPrint software developed by Wageningen University and Research, Wageningen, The Netherlands [21]. The FeedPrint calculates the CFP of feed raw materials during their complete life cycle and it has been developed to gain insight into GHG emissions during the production and utilization chain of feed and to identify mitigation options. Supplementary Table S1 presents the CFP of the common feed raw materials used for the estimation of feed CFP. The contribution of the feed raw materials to the feed CFP was estimated by multiplying the inclusion level of the raw material and the CFP per kg of raw material ($g\ CO_2$-eq/kg). The average feed CFP was expressed as $g\ CO_2$-eq/kg diet. Figure 1 shows that replacing SBM (0% SRU) with 1% and 3% SRU in the diets of growing bulls decreased calculated feed CFP by 18% and 54%, respectively.

2.4. Statistical Analysis

For experiment 1 (in vitro fermentation study), the experimental unit was the individual fermentation vessel. For each of the two diets (high-forage or high-grain), there were ten experimental groups: a control group and three nitrogen supplements (SBM, SRU and urea) each at three levels of nitrogen (low, medium or high). All data were subjected to one-way ANOVA as a factorial arrangement, to determine the effect of dietary treatment on rumen fermentation parameters. Tukey's HSD test was used for multiple comparisons of treatment means when significance was detected at $p < 0.05$ and a tendency for treatment effect was observed when $0.05 < p \leq 0.10$. Statistical analyses were performed using Minitab® software (Minitab, v18, State College, TX, USA). Data from experiment 2 (in situ degradability study) were analyzed as a linear mixed model to test the effect of the diets as a fixed factor and the cow was considered as a random term using SAS® software. Diet effect was declared significant when $p < 0.05$. For experiment 3 (in vivo study), feed intake data were analyzed using a linear mixed-effects model with repeated measurements using SAS® software. The model accounts for the fixed effects of treatment, time and the interaction between these two factors, plus the random effect of the pen as the experimental unit and the initial body weight as a covariate. Blood parameters at 42 d were analyzed using a linear model with the fixed effects of treatment including data of day 0 as covariate plus the random effect of the pen. Tukey's HSD test was used for multiple comparisons of treatment groups when a significant effect of the treatment was found at $p < 0.05$ for growth performance parameters and $p \leq 0.10$ for blood indices.

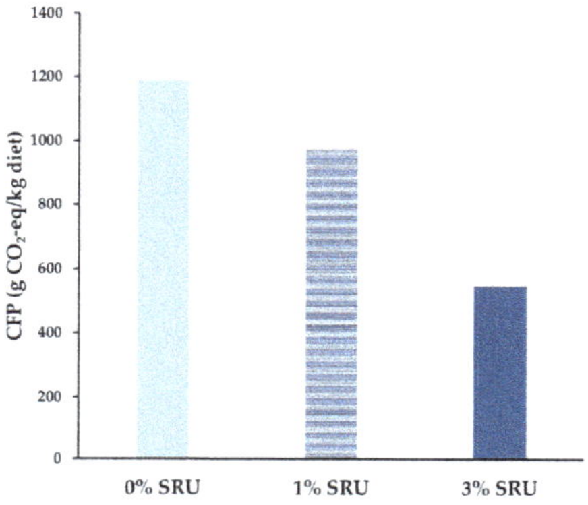

Figure 1. Calculated carbon footprint of concentrate diets of growing cattle when soybean meal (0% SRU) was replaced with increasing levels of slow-release urea (1% SRU and 3% SRU).

3. Results

3.1. Experiment 1: In Vitro Rumen Fermentation

Although the effects of supplementing different nitrogen sources (SBM, SRU and urea) on in vitro rumen fermentation in high-forage and high-grain diets were measured over 24 h of incubation, the most discerning effects were obtained at 9 h incubation. Thus, the effects of treatments on in vitro fermentation characteristics at 9 h are reported here (Figures 2–7) while results of total gas production, short-chain fatty acids, pH, ammonia and lactate concentration measured at all time points (0 h, 9 h and 12 h of incubation) are presented in Tables S2–S15. Total gas production at 9 h ($p > 0.05$) was unaffected when different nitrogen sources (SBM, SRU and urea) were supplemented with high-forage diets (Figure 2A). However, total gas production at 9 h for the high-grain diets increased significantly ($p < 0.05$) for both urea and SRU treatments in a dose-dependent manner while the isonitrogenous amount of SBM did not affect total gas production (Figure 2B). The addition of urea significantly increased ($p < 0.05$) ammonia concentration in high-forage diets after 9 h, whereas there was no difference ($p > 0.05$) between SRU, SBM and control treatments (Figure 3A). For high-grain diets, only the highest dose (i.e., 5.35 mg N) of urea increased ammonia concentration ($p < 0.05$) after 9 h, with no difference ($p > 0.05$) observed between SRU, SBM and control treatments (Figure 3B). Nitrogen supplements in high-forage diets did not influence the pH of the fermentation vessels (Figure 4A) whereas medium and high dosage (i.e., 3.22 mg N and 5.35 mg N) of SRU and urea in high-grain diets lowered the pH of the fermentation vessels (Figure 4B). Compared with the control treatment, lactic acid accumulation at 9 h was not affected ($p > 0.05$) by nitrogen supplements (SBM, SRU and urea) in high-forage diets (Figure 5A) whereas high dosage of SRU and urea treatments (medium and high dosages) increased lactic acid concentration in high-grain diets (Figure 5B). Notably, SRU resulted in 21 to 33% lower

($p < 0.05$) accumulation of lactic acid in the high-grain diets compared to urea (Figure 5B). However, the residual concentration of lactic acid was marginal at 12 h with the high-forage diet while its concentration at 12 h was still significant with the high-grain diet (Table S14). Total volatile fatty acids concentration measured at 9 h was not different ($p > 0.05$) between control, SBM, SRU and urea treatments in both high-forage and high-grain diets (Figure 6A,B). The effect of nitrogen supplements on bacterial density was observed ($p < 0.05$) in high-forage diets (Figure 7A) but not ($p > 0.05$) in high-grain diets (Figure 7B). In high-forage diets, both SBM and urea inclusion slightly increased the total bacterial density at this single analyzed time point (12 h). In general, the high-grain diet yielded higher bacterial density than the high-forage diet.

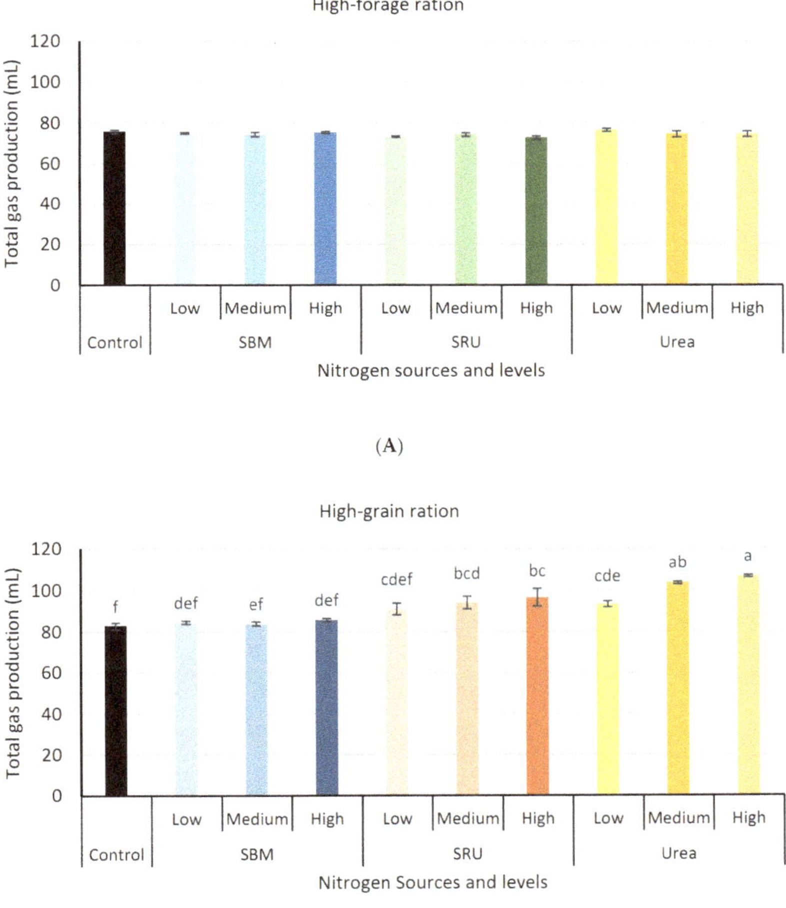

(**A**)

(**B**)

Figure 2. Total gas production measured at 9 h in an in vitro fermentation model incubated with (**A**) high-forage (20:80, wheat: grass silage) ration and (**B**) high-grain ration (60:40, wheat: grass silage) supplemented with three different nitrogen sources (soybean meal (SBM), slow-release urea (SRU) and urea) at increasing nitrogen levels (low, medium and high). Low, medium and high nitrogen levels correspond to 1.07 mg N, 3.22 mg N and 5.35 mg N, respectively. Values are presented as means with standard error bars. [a–f] Bars with different letters are significantly different ($p < 0.05$).

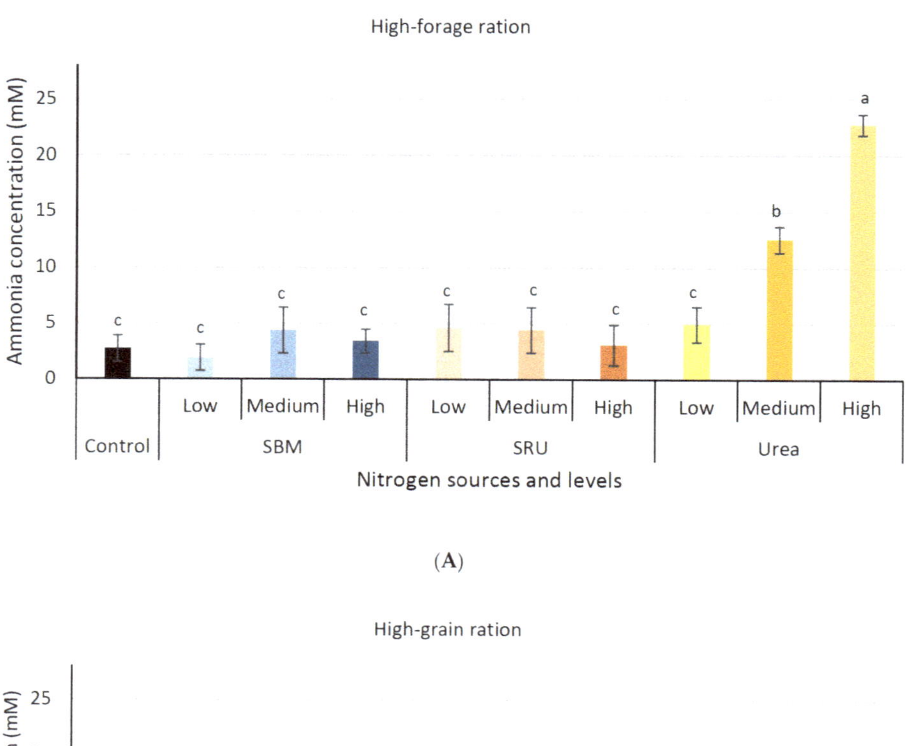

(**A**)

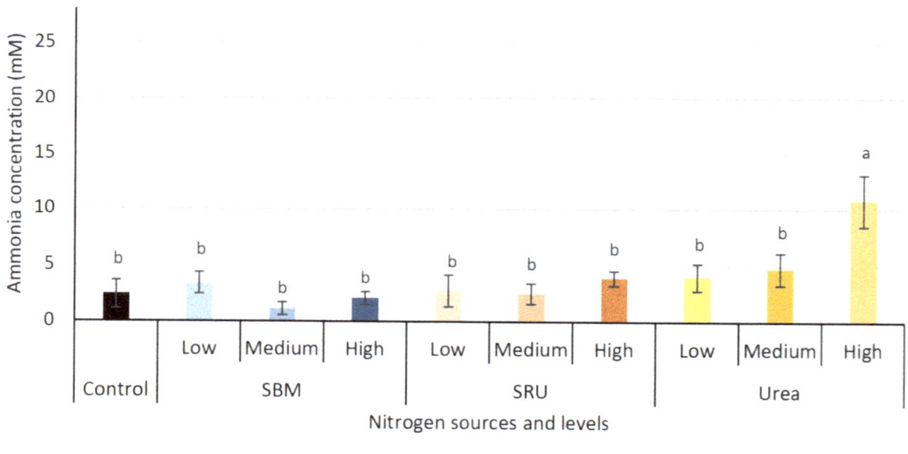

(**B**)

Figure 3. Ammonia (NH_3) concentration measured at 9 h in an in vitro fermentation model incubated with (**A**) high-forage (20:80, wheat: grass silage) ration and (**B**) high-grain ration (60:40, wheat: grass silage) supplemented with three different nitrogen sources (soybean meal (SBM), slow-release urea (SRU) and urea) at increasing nitrogen levels (low, medium and high). Low, medium and high nitrogen levels correspond to 1.07 mg N, 3.22 mg N and 5.35 mg N, respectively. Values are presented as means with standard error bars. [a–c] Bars with different letters are significantly different ($p < 0.05$).

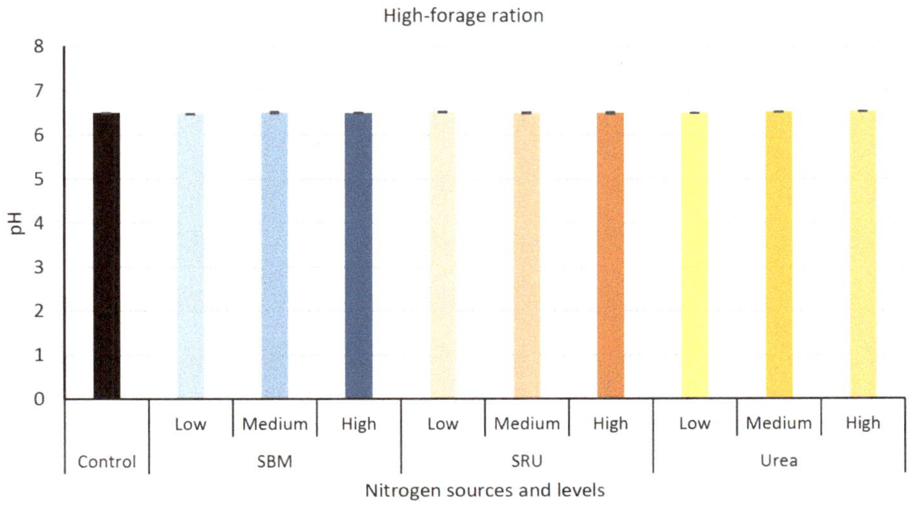

(**A**)

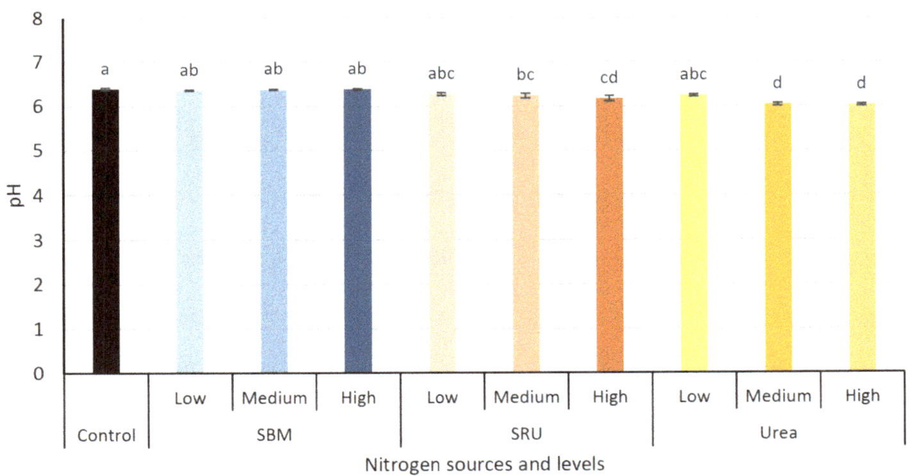

(**B**)

Figure 4. pH measured at 9 h in an in vitro fermentation model incubated with (**A**) high-forage (20:80, wheat: grass silage) ration and (**B**) high-grain ration (60:40, wheat: grass silage) supplemented with three different nitrogen sources (soybean meal (SBM), slow-release urea (SRU) and urea) at increasing nitrogen levels (low, medium and high). Low, medium and high nitrogen levels correspond to 1.07 mg N, 3.22 mg N and 5.35 mg N, respectively. Values are presented as means with standard error bars. [a–d] Bars with different letters are significantly different ($p < 0.05$).

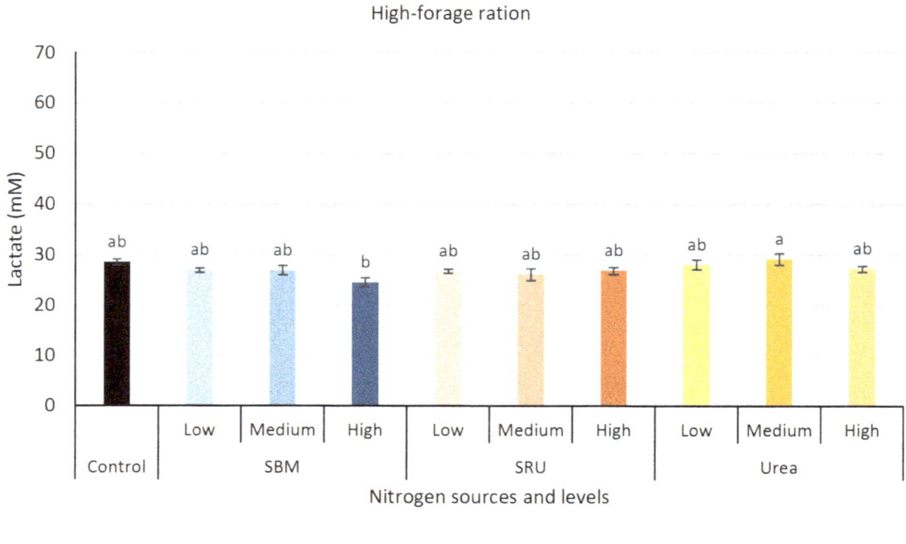

(**A**)

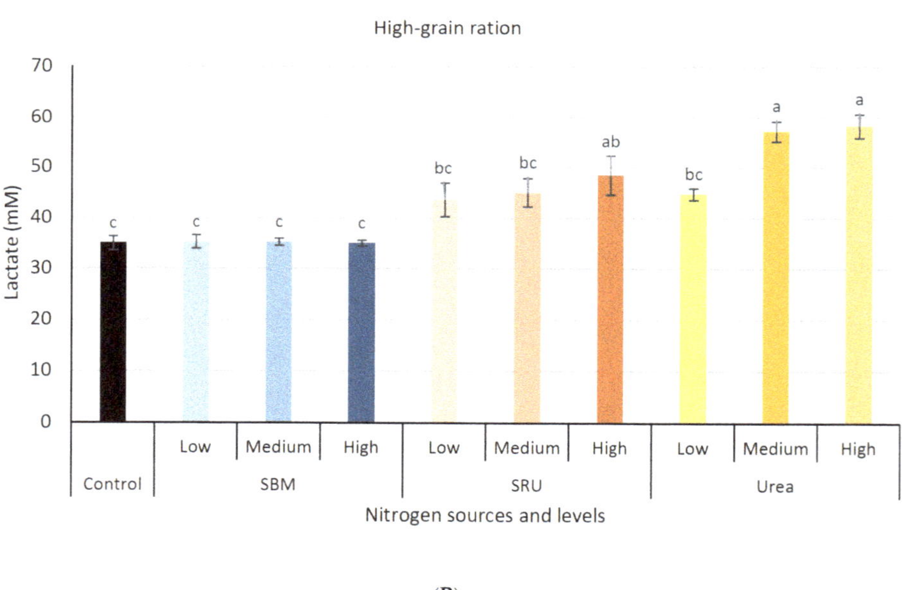

(**B**)

Figure 5. Lactate concentration measured at 9 h in an in vitro fermentation model incubated with (**A**) high-forage (20:80, wheat: grass silage) ration and (**B**) high-grain ration (60:40, wheat: grass silage) supplemented with three different nitrogen sources (soybean meal (SBM), slow-release urea (SRU) and urea) at increasing nitrogen levels (low, medium and high). Low, medium and high nitrogen levels correspond to 1.07 mg N, 3.22 mg N and 5.35 mg N, respectively. Values are presented as means with standard error bars. [a–c] Bars with different letters are significantly different ($p < 0.05$).

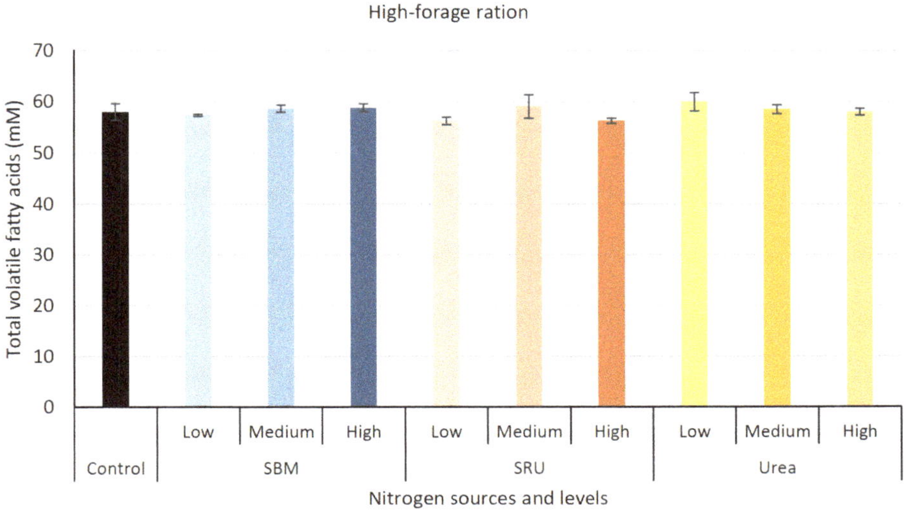

(**A**)

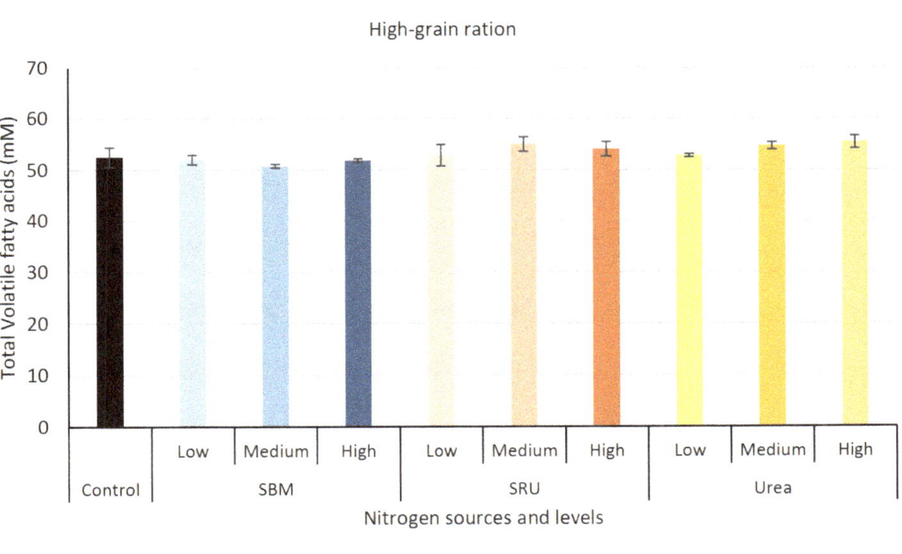

(**B**)

Figure 6. Total volatile fatty acids concentration measured at 9 h in an in vitro fermentation model incubated with (**A**) high-forage (20:80, wheat: grass silage) ration and (**B**) high-grain ration (60:40, wheat: grass silage) supplemented with three different nitrogen sources (soybean meal (SBM), slow-release urea (SRU) and urea) at increasing nitrogen levels (low, medium and high). Low, medium and high nitrogen levels correspond to 1.07 mg N, 3.22 mg N and 5.35 mg N, respectively. Values are presented as means with standard error bars.

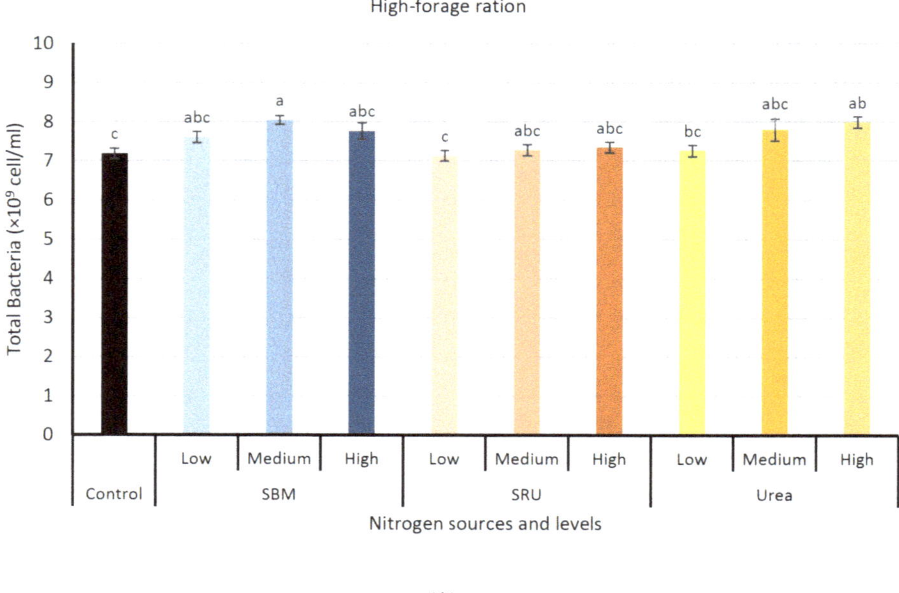

(**A**)

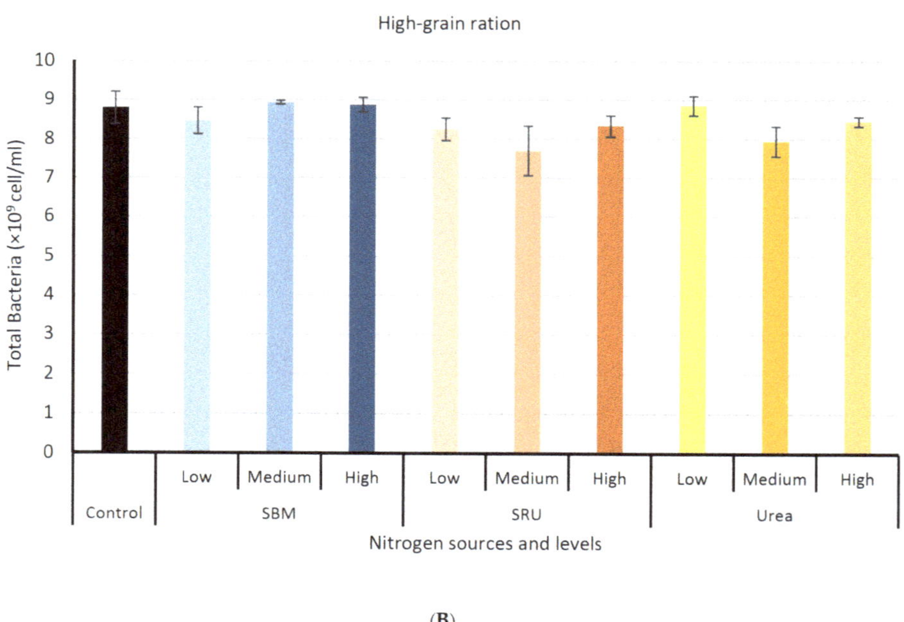

(**B**)

Figure 7. Total bacteria concentration measured at 12 h in an in vitro fermentation model incubated with (**A**) high-forage (20:80, wheat: grass silage) ration and (**B**) high-grain ration (60:40, wheat: grass silage) supplemented with three different nitrogen sources (soybean meal (SBM), slow-release urea (SRU) and urea) at increasing nitrogen levels (low, medium and high). Low, medium and high nitrogen levels correspond to 1.07 mg N, 3.22 mg N and 5.35 mg N, respectively. Values are presented as means with standard error bars. [a–c] Bars with different letters are significantly different ($p < 0.05$).

3.2. Experiment 2: In Situ Degradability

The effects of replacing one-third of the SBM in the control diet with SRU on in situ dry matter and nitrogen degradability estimates are presented in Table 4. The potentially degradable DM fraction was lower ($p < 0.05$) in the SRU diet whereas other estimated parameters (c, K_d and ERD) were not affected. Similarly, the replacement of SBM with SRU did not significantly affect ($p > 0.05$) the nitrogen degradability estimates. However, ERD of dietary nitrogen tended ($p = 0.052$) to be lower in the SRU diet compared to the control diet.

Table 4. Effect of replacing soybean meal with slow-release urea (SRU) on estimated parameters of in situ dry matter (DM) and nitrogen degradation in the rumen.

Parameter	Control	SRU	SEM	*p*-Value
		DM degradation		
b	68.30	62.31	0.029	0.004
c	29.05	31.27	0.686	0.263
K_d	0.15	0.16	0.006	0.640
ERD	51.86	51.71	0.219	0.716
		Nitrogen degradation		
b	88.40	85.91	0.509	0.179
c	11.60	14.09	0.509	0.179
K_d	0.10	0.10	0.002	0.252
ERD	54.40	53.98	0.024	0.052

b: potentially degradable fraction; c: undegradable fraction; K_d: fractional rate of degradation; ERD: effective rumen degradability. SEM: standard error of the mean.

3.3. Experiment 3: Effect on Feed Intake and Blood Indices in Growing Cattle

There was no effect ($p > 0.05$) of dietary treatment on dry matter intake (5.06, 4.89 and 5.01 kg DM/day for 0%, 1% and 3% SRU respectively; $p = 0.721$).

The blood hematologic indices for all dietary treatments were within normal physiological ranges (Table 5). Compared with 0% SRU (control diet), bulls fed SRU at 1% or 3% diets had lower ($p < 0.05$) hemoglobin and packed cell volume. Bulls fed 3% SRU diet had lower mean corpuscular volume ($p = 0.090$) and mean corpuscular hemoglobin ($p = 0.061$) compared to those fed 0% SRU diet. Mean corpuscular hemoglobin concentrations (39.4–39.6 g/dL) were slightly higher than reference values (30–38 g/dL [33]) but did not differ between the control and SRU diets.

Table 5. Effect of slow-release urea (SRU) supplementation in concentrate diets on hematological indices of growing cattle.

Parameter	0% SRU	1% SRU	3% SRU	SEM	*p*-Value	Reference Values [1]
Red cell count ($\times 10^6$ cells/μL)	9.6	9.5	9.7	0.15	0.649	6.5–11.9
Hemoglobin (g/dL)	13.0 [a]	12.3 [b]	12.3 [b]	0.19	0.028	8–14.1
Packed cell volume (%)	33.1 [a]	31.1 [b]	31.2 [b]	0.44	0.007	23–42
Mean corpuscular volume (fL)	34.2 [a]	33.1 [ab]	32.6 [b]	0.44	0.090	26.6–44.3
Mean corpuscular hemoglobin (pg)	13.5 [a]	13.1 [ab]	12.8 [b]	0.18	0.061	9.1–15.6
White cell count ($\times 10^3$ cells/μL)	11.6	10.9	11.9	0.92	0.609	5.6–13.7
Neutrophils ($\times 10^3$ cells/μL)	4.6	3.7	4.5	0.67	0.428	0.6–6.1
Lymphocytes ($\times 10^3$ cells/μL)	6.1	6.4	6.8	0.46	0.527	2.2–8.7
Monocytes ($\times 10^3$ cells/μL)	0.5	0.5	0.5	0.06	0.809	0.1–0.2
Eosinophils ($\times 10^3$ cells/μL)	0.2	0.2	0.2	0.08	0.902	0–0.3
Basophils ($\times 10^3$ cells/μL)	0.3	0.3	0.4	0.03	0.519	0–0.1
Platelets ($\times 10^3$ cells/μL)	580	591	524	48	0.752	220–950

SEM: Standard error of the mean. [1] Reference values were according to Lumsden, et al. [33], Martin and Lumsden [34], and Mohri, et al. [35]. [ab] Values in the same row with different superscripts are significantly different ($p \leq 0.10$).

Table 6 depicts the blood biochemical indices in growing bulls fed concentrate diets in which SBM was replaced with 1% or 3% SRU. Albumin concentrations differed ($p < 0.05$) between treatments, but all values were within the normal physiological range. Total protein values were significantly lower ($p < 0.05$) in bulls fed diets containing SRU compared to bulls fed the control diet and were slightly under the reference values. The activity of aspartate aminotransferase was lower ($p = 0.07$) in bulls fed 1% SRU compared with those fed 0% SRU, whereas a further reduction in aspartate aminotransferase activity was not observed at 3% SRU. Serum ammonia concentrations were below the toxic limit (i.e., 5 mg/dL) and did not differ ($p > 0.05$) between dietary treatments.

Table 6. Effect of slow-release urea (SRU) supplementation in concentrate diet on serum biochemical indices of growing cattle.

Parameter	0% SRU	1% SRU	3% SRU	SEM	p-Value	Reference Values [1]
Glucose (mg/dL)	101.2	103.0	102.9	2.91	0.879	45–135
Aspartate aminotransferase (IU/L)	66.1 [a]	56.9 [b]	62.5 [ab]	3.03	0.067	<60
Alanine aminotransferase (IU/L)	17.2	14.9	16.6	0.75	0.114	3–18
Gamma-glutamyl transferase (IU/L)	13.4	12.1	11.7	1.13	0.534	<16
Urea (mg/dL)	27.5	28.2	24.4	1.44	0.180	21–54
Uric acid (mg/dL)	0.4	0.4	0.4	-	0.679	-
Albumin (g/dL)	3.7 [a]	3.4 [c]	3.6 [b]	0.03	0.001	2.5–3.8
Ammonia (mg/dL)	0.7	0.7	0.9	0.09	0.105	<5
Total protein (g/dL)	6.8 [a]	6.6 [b]	6.5 [b]	0.09	0.031	6.8–7.5

SEM: standard error of the mean. [1] Reference values were according to EFSA [14], Lumsden, et al. [33], Martin and Lumsden [34], and Mohri, et al. [35]. [a–c] Values in the same row with different superscripts are significantly different ($p \leq 0.10$).

4. Discussion

Vegetable protein sources, such as SBM and rapeseed meal, are commonly utilized to supply RDP in ruminant diets. Emissions from ruminant feeds represent 36% of the GHG emissions attributed to beef and milk production [3]. Thus, formulating environmentally friendly ruminant feeds offers an opportunity to mitigate the negative impacts of beef and milk production on GHG emissions and climate change [7]. Vegetable protein sources could constitute high CFP due to the impact of cultivation on land-use changes. This has increased the interest in utilizing NPN sources for providing RDP in ruminant nutrition. In the experimental design of the in vivo study (experiment 3), we employed a diet reformulation strategy that reduced the feed CFP by using the inclusion of 1% or 3% SRU for isonitrogenous replacement of SBM in concentrate diets of growing beef cattle. Consistent with our experimental design, Reddy, et al. [10] reported that feed CFP was decreased by 23% when SBM was replaced by up to 1.2% SRU in a total mixed ration of sheep. Similarly, replacing cottonseed meal with SRU at a 2% level of concentrate diet reduced the feed CFP of dairy buffalo by 25% [9]. Moreover, a recent meta-analysis study revealed that the replacement of vegetable protein sources with an average inclusion of 0.58% DM SRU reduced the CFP of feed use for milk production by 14.5% [36]. In addition to a reduction in feed CFP, these previous studies showed that feeding SRU to ruminants could offer other environmental benefits such as lower global warming potential for meat and milk production, lower eutrophication potential and reduction of land and virtual water requirement for feed. Overall, this information suggests that SRU could be an eco-friendly alternative to vegetable protein sources in ruminant diets.

As mentioned previously, feed-grade urea is a commonly used NPN source in ruminant diets, but its utilization is limited due to rapid degradation in the rumen, producing excess ammonia that can increase the risk of ammonia toxicity on animal health and increase nitrogen excretion to the environment [11,37]. Alternatively, coating technologies have been used to develop SRU with a reduced rate of hydrolysis in the rumen, increasing the synchronization of ammonia and fermentable carbohydrate for microbial protein synthesis in the rumen. This reduced rate of hydrolyses in the rumen may reduce their toxicity when supplemented at greater rates compared with feed-grade urea, increasing their potential to substitute vegetal protein sources and therefore reducing the CFP.

In the first study, the effect of energy supply and N doses was evaluated on microbial fermentation in vitro. Microbial fermentation of feed substrates in the rumen produces VFA and microbial proteins that supply ruminant animals with energy and highly digestible proteins used for maintenance and production purposes [38]. Thus, strategies aimed at reformulating ruminant diets with SRU must not impair ruminal fermentation to avoid negative effects on animal performance. It is noteworthy that the most discerning effects of supplementing high-forage and high-grain diets with different nitrogen sources (SBM, SRU and urea) on rumen fermentation were observed at 9 h incubation in the fermentation vessels. Thus, the effect of treatments on in vitro fermentation characteristics at 9 h were reported in this study. In general, the present results indicated that supplementation of SRU in high-forage and high-grain cattle rations did not impair in vitro ruminal fermentation characteristics. The effect of SRU on ruminal fermentation was similar to that of SBM when supplemented in high-forage and high-grain cattle diets. Gas production in the rumen directly results from the microbial digestion of feed substrates and indirectly from buffering of acids generated as a result of fermentation [39]. Menke [40] indicated that the amount of gas production reflects the extent and rate of digestion of soluble and insoluble carbohydrates and the production of VFA. In this study, higher gas production was obtained from the incubation of high-grain diets compared to high-forage diets, possibly due to greater fermentation of higher non-fibre carbohydrate levels in the high-grain diet [39]. Dose-dependent stimulation of gas production by SRU and urea, when supplemented in high-grain diets, could suggest that nitrogen supply may have been limiting rumen fermentation. In contrast to the effect of NPN sources (SRU and urea), SBM supplementation did not increase in vitro gas production, possibly indicating that more soluble protein might be required to increase the fermentation of the high-grain diets. However, the effect of NPN sources on increasing total gas production of high-grain diets did not result in greater total VFA concentration.

Furthermore, the current results indicated that SRU resulted in similar ammonia accumulation as SBM but lower than that of urea, confirming the reduced hydrolysis of SRU in the simulated rumen environment. This effect could be explained by the similarity in the nitrogen disappearance of SRU and that of SBM in the rumen, which is slower than that of feed-grade urea [41]. This assertion is further supported by the similarity in the in situ nitrogen degradation estimates when one-third of SBM was replaced by SRU in diets incubated in the rumen of cows reported herein. In further agreement with our observation, Garcia-Gonzalez, et al. [19] demonstrated that the dissolution rate of SRU in the rumen was lower than that of urea, resulting in lower ruminal and blood ammonia concentrations in steers. This implies that reduced ruminal hydrolysis of SRU produces lower ammonia concentration that could decrease the risk of ammonia toxicity and improve nitrogen utilization efficiency through better synchronization of ammonia with available fermentable energy in the rumen [11]. An increase in rumen pH is a major contributing factor to the toxicity of urea, as the permeability of the rumen epithelium to ammonia increases as rumen pH increases [42]. In this study, there was no effect of supplementing SRU or urea in high-forage diets on the pH of the fermentation vessels. This is in agreement with previous observations reported when urea or SRU was supplemented in high-fibre diets similar to the high-forage diets incubated herein [17,43]. However, supplementing medium and high dosage (i.e., 3.22 mg N and 5.35 mg N) of SRU and urea in high-grain diets lowered the pH of the fermentation vessels. This observation is contrary to the assertion that ruminal ammonia from urea could increase ruminal pH, as ammonia protonates to ammonium [44]. Notably, the lower pH in the present study was accompanied by an increase in lactic acid accumulation particularly in high-grain diets supplemented with NPN sources (SRU and urea). The high-grain diet presumably induced an acidotic condition in the ruminal incubation system, typical of the effect of high concentrate diets [45]. It is well documented that rapid fermentation of starch in high concentrate diets results in the accumulation of lactic acid, which induces lower ruminal pH [45]. Thus, the lower pH found in our study could be attributed to the acidotic potential of the high-grain diet, but NPN sources appeared to stimulate this acidotic condition to a lesser extent for SRU compared to urea. It is noteworthy that the decrease in ruminal pH was not as dramatic as the elevated level of lactic acid in the urea-based treatments, which could be related to the counteracting effect of ammonia accumulation on ruminal pH.

It is crucial to consider that reformulating diets with SRU does not impair ruminant health and performance due to their direct relationship with animal welfare and farm profitability [46,47]. In this regard, we examined the tolerance of growing cattle to feeding a high inclusion level of SRU at 3% of the concentrate diet. It is noteworthy that our feeding trial was not primarily designed to test the effect of SRU in improving the production performance of cattle. Growth performance parameters are not discussed in the present paper and only intake data are considered to analyze if SRU exhibits toxicity potential that affects animal health and consequently affects the production performance of cattle. Replacement of SBM with 1% or 3% SRU in concentrate diets did not affect the intake or health of growing bulls, suggesting that feeding up to 3% SRU was well tolerated by cattle. In agreement with this result, Bourg, et al. [48] have also shown that the inclusion of 3.1% DM of SRU in concentrate diet did not negatively affect the growth performance and carcass characteristics of steers. The lack of negative effect of SRU on animal performance is consistent with results of in vitro and in situ experiments which indicated that SRU does not impair rumen fermentation and feed degradability. Similarly, previous studies have shown that dietary inclusion up to 1.75% DM SRU did not affect the growth performance of cattle [16,47,48]. However, other studies have demonstrated positive effects of dietary SRU on feed efficiency of beef and dairy cattle production [48,49]. A recent meta-analysis showed that partial replacement of vegetable protein sources with an average SRU inclusion of 0.88% DM diet improved the liveweight gain and feed efficiency of growing and finishing beef cattle [24]. The authors identified different diet- and production-related factors that contributed to variations in the performance response of cattle when SRU is supplemented. Dietary inclusion of corn silage as the forage source was particularly found to improve liveweight gain and feed efficiency while other roughage sources did not affect these performance parameters. This positive effect was explained by the high digestible energy and digestibility value of corn silage which could have provided fermentable carbohydrate with better synchronization with ammonia to optimize microbial protein synthesis [24]. In contrast, straw was fed as the roughage source in the present study, which might have partly contributed to the lack of positive effect of SRU on the growth performance of beef cattle.

Furthermore, the potential of SRU supplementation to cause ammonia toxicity in cattle was determined by measuring blood hematological and biochemical parameters [50,51]. Excess ammonia produced from rapid ruminal hydrolysis of RDP sources, such as urea, above the requirement of microbes may be absorbed across the rumen epithelium and converted to urea in the liver through hepatic detoxification [52]. The liver can remove ammonia added to the portal blood up to a maximum of 182 mg/min before peripheral blood concentration increases [53]. The accumulation of ammonia in the blood of urea-fed cattle is the primary cause of urea toxicity [42]. Serum ammonia concentration is one of the primary biomarkers used to evaluate urea toxicity [54]. In the current study, serum ammonia concentrations (0.7–0.9 mg/dL) measured at 1 h post-feeding were below the toxic limit (i.e., 5 mg/dL) and did not differ between the three dietary treatments. This observation is consistent with the lack of treatment effect on plasma ammonia concentrations when 0.55% SRU or 0.50% feed-grade urea were fed to dairy cows [18]. Similarly, the serum urea concentration measured in this study was not different between treatments and was within recommended range. Holder, et al. [55] showed that rumen ammonia and plasma urea were higher in steers fed feed-grade urea compared to those fed SRU at two levels of dietary CP (12.1% and 10.9%), suggesting that SRU offers a safer alternative to feed-grade urea. Garcia-Gonzalez, et al. [19] also reported that replacing feed-grade urea with SRU resulted in reduced rumen ammonia concentration and did not increase postprandial plasma urea concentration. Moreover, aspartate aminotransferase (AST), alanine aminotransferase

(ALT) and gamma-glutamyl transferase (GGT) are key enzymes that reflect the condition of liver function and they are associated with the welfare of animals [50,56]. Chronic urea toxicity could result in intense hepatic nitrogen metabolism, which may be manifested by higher serum activity of AST, ALT and GGT enzymes in ruminants [50,57]. Thus, the lack of substantial increase in these key enzymes is further evidence that feeding 3% SRU did not induce toxicity in growing bulls. Overall, blood hematological and biochemical indices measured in this study were within the normal physiological ranges for healthy cows, suggesting that feeding high inclusion level of up to 3% SRU in a concentrate diet was well tolerated by the bulls and did not exhibit a negative effect on animal health.

5. Conclusions

These results demonstrated that SRU is a viable NPN source that can be utilized to partially replace vegetable protein sources, such as SBM, without negative impact on in vitro rumen fermentation and in situ feed degradability under the current experimental conditions. Replacement of vegetable protein sources with SRU could reduce the CFP of ruminant diets, thereby contributing to lower environmental impacts of ruminant milk and meat production. Moreover, dietary inclusion of SRU can be tolerated at up to 3% in cattle diets without negative impacts on feed intake and health indices of bulls.

Supplementary Materials: The following are available online at https://www.mdpi.com/2071-1050/13/5/2464/s1. Table S1. Carbon footprint (CFP) of common feed raw materials used for calculating the CFP of diets used in experiment 3. Table S2. Total gas production (mL) measured in an in vitro fermentation model. Table S3. Acetate concentration (mM) measured in an in vitro fermentation model. Table S4. Propionate concentration (mM) measured in an in vitro fermentation model. Table S5. Isobutyric acid concentration (mM) measured in an in vitro fermentation model. Table S6. Butyrate concentration (mM) measured in an in vitro fermentation model. Table S7. Total short-chain fatty acids concentration (mM) measured in an in vitro fermentation model. Table S8. Isovaleric acid concentration (mM) measured in an in vitro fermentation model. Table S9. Valeric acid concentration (mM) measured in an in vitro fermentation model. Table S10. Branched volatile fatty acids concentration (mM) measured in an in vitro fermentation model. Table S11. 2-methylbutyric acid concentration (mM) measured in an in vitro fermentation model. Table S12. pH measured in an in vitro fermentation model. Table S13. Ammonia concentration (mM) measured in an in vitro fermentation model. Table S14. Lactate concentration (mM) measured in an in vitro fermentation model. Table S15. Total volatile fatty acids concentration (mM) measured in an in vitro fermentation model.

Author Contributions: Conceptualization, J.A., M.D. and C.A.M.; methodology, J.A., M.D.; formal analysis, J.A., M.D. and S.A.S.; investigation, J.A., M.D., S.S.; resources, J.A., M.D.; data curation, J.A., M.D.; writing—original draft preparation, V.H. and S.A.S.; writing—review and editing, J.A., M.D., J.D.K., S.A.S. and C.A.M.; visualization, V.H. and S.A.S.; supervision, J.A., M.D. and C.A.M.; project administration, C.A.M.; funding acquisition, C.A.M. All authors have read and agreed to the published version of the manuscript.

Funding: The research was financially supported by Alltech SARL (France), which also provided the test products used in this study.

Institutional Review Board Statement: The animals used as a source of rumen fluid or *in situ* trial were cannulated and maintained in the research facility of Alimetrics Ltd. in Southern Finland, following the European Union (EU) Directive 2010/63/EU on the protection of animals used for experimental or other scientific purposes. The cannulation was approved by the Animal Experiment Board in Finland. Additionally, the research protocol of the animal feeding trial followed the EU Directive 2010/63/EU and the animals were managed according to the regulations of the Animal Care Committee of the Institute of Agrifood Research and Technology, Spain.

Informed Consent Statement: Not applicable.

Data Availability Statement: Data will be provided upon request.

Acknowledgments: The authors thank all members of the IRTA Ruminant Production team and the farm team of BonArea Agrupa for the hard work and enthusiasm, and thanks to the support of the Generalitat de Catalunya through the CERCA Programme. The authors appreciate Kamal Mjoun (Alltech Inc., Nicholasville, KY, USA) for his expert assistance with the data analysis. C.A.M. wishes to acknowledge the late Pearse Lyons (Alltech Inc., Nicholasville, KY, USA) for his leadership in sustainable agriculture.

Conflicts of Interest: The authors S.A.S., V.H., J.D.K. and C.A.M. are employees of Alltech which produces and markets Optigen®, the commercial slow-release urea evaluated in this study. The research was funded by Alltech SARL (France).

References

1. Adesogan, A.T.; Havelaar, A.H.; McKune, S.L.; Eilittä, M.; Dahl, G.E. Animal source foods: Sustainability problem or malnutrition and sustainability solution? Perspective matters. *Glob. Food Secur.* **2020**, *25*, 100325. [CrossRef]
2. Herrero, M.; Thornton, P.K. Livestock and global change: Emerging issues for sustainable food systems. *Proc. Natl. Acad. Sci. USA* **2013**, *110*, 20878–20881. [CrossRef] [PubMed]
3. Gerber, P.J.; Steinfeld, H.; Henderson, B.; Mottet, A.; Opio, C.; Dijkman, J.; Falcucci, A.; Tempio, G. *Tackling Climate Change through Livestock: A Global Assessment of Emissions and Mitigation Opportunities*; Food and Agriculture Organization of the United Nations (FAO): Rome, Italy, 2013.
4. Wattiaux, M.; Uddin, M.; Letelier, P.; Jackson, R.; Larson, R. Invited Review: Emission and mitigation of greenhouse gases from dairy farms: The cow, the manure, and the field. *Appl. Anim. Sci.* **2019**, *35*, 238–254. [CrossRef]
5. Wilkinson, J.; Garnsworthy, P. Impact of diet and fertility on greenhouse gas emissions and nitrogen efficiency of milk production. *Livestock* **2017**, *22*, 140–144. [CrossRef]
6. Opio, C.; Gerber, P.; Mottet, A.; Falcucci, A.; Tempio, G.; MacLeod, M.; Vellinga, T.; Henderson, B.; Steinfeld, H. *Greenhouse Gas Emissions from Ruminant Supply Chains–A Global Life Cycle Assessment*; Food and Agriculture Organization of the United Nations: Rome, Italy, 2013.
7. Wilkinson, J.; Garnsworthy, P. Dietary options to reduce the environmental impact of milk production. *J. Agric. Sci.* **2017**, *155*, 334–347. [CrossRef]
8. Salami, S.; Luciano, G.; O'Grady, M.; Biondi, L.; Newbold, C.; Kerry, J.; Priolo, A. Sustainability of feeding plant by-products: A review of the implications for ruminant meat production. *Anim. Feed Sci. Technol.* **2019**, *251*, 37–55. [CrossRef]
9. Reddy, P.R.K.; Kumar, D.S.; Rao, E.R.; Seshiah, C.V.; Sateesh, K.; Rao, K.A.; Reddy, Y.P.K.; Hyder, I. Environmental sustainability assessment of tropical dairy buffalo farming vis-a-vis sustainable feed replacement strategy. *Sci. Rep.* **2019**, *9*, 1–16. [CrossRef]
10. Reddy, P.R.K.; Kumar, D.S.; Rao, E.R.; Seshiah, C.V.; Sateesh, K.; Reddy, Y.P.K.; Hyder, I. Assessment of eco-sustainability vis-à-vis zoo-technical attributes of soybean meal (SBM) replacement with varying levels of coated urea in Nellore sheep (Ovis aries). *PLoS ONE* **2019**, *14*, e0220252.
11. Cherdthong, A.; Wanapat, M. Development of urea products as rumen slow-release feed for ruminant production: A review. *Aust. J. Basic Appl. Sci.* **2010**, *4*, 2232–2241.
12. Calsamiglia, S.; Ferret, A.; Reynolds, C.; Kristensen, N.B.; Van Vuuren, A. Strategies for optimizing nitrogen use by ruminants. *Animal* **2010**, *4*, 1184–1196. [CrossRef] [PubMed]
13. Cope, R.B. Nonprotein nitrogen (urea) and hyperammonemia. In *Veterinary Toxicology*; Elsevier: Amsterdam, The Netherlands, 2018; pp. 1093–1097.
14. EFSA. EFSA Panel on Additives Products or Substances used in Animal Feed. Scientific Opinion on the safety and efficacy of Urea for ruminants. *EFSA J.* **2012**, *10*, 2624.
15. Owens, F.; Lusby, K.; Mizwicki, K.; Forero, O. Slow ammonia release from urea: Rumen and metabolism studies. *J. Anim. Sci.* **1980**, *50*, 527–531. [CrossRef]
16. Tedeschi, L.; Baker, M.; Ketchen, D.; Fox, D. Performance of growing and finishing cattle supplemented with a slow-rlease urea product and urea. *Can. J. Anim. Sci.* **2002**, *82*, 567–573. [CrossRef]
17. Taylor-Edwards, C.; Hibbard, G.; Kitts, S.; McLeod, K.; Axe, D.; Vanzant, E.; Kristensen, N.; Harmon, D. Effects of slow-release urea on ruminal digesta characteristics and growth performance in beef steers. *J. Anim. Sci.* **2009**, *87*, 200–208. [CrossRef] [PubMed]
18. Sinclair, L.; Blake, C.; Griffin, P.; Jones, G. The partial replacement of soyabean meal and rapeseed meal with feed grade urea or a slow-release urea and its effect on the performance, metabolism and digestibility in dairy cows. *Animal* **2012**, *6*, 920–927. [CrossRef] [PubMed]
19. Garcia-Gonzalez, R.; Tricarico, J.; Harrison, G.; Meyer, M.; McLeod, K.; Harmon, D.; Dawson, K. Optigen®is a sustained release source of non-protein nitrogen in the rumen. *J. Anim. Sci.* **2007**, *85*, 98.
20. Ribeiro, S.; Vasconcelos, J.; Morais, M.; Ítavo, C.; Franco, G. Effects of ruminal infusion of a slow-release polymer-coated urea or conventional urea on apparent nutrient digestibility, in situ degradability, and rumen parameters in cattle fed low-quality hay. *Anim. Feed Sci. Technol.* **2011**, *164*, 53–61. [CrossRef]

21. Vellinga, T.V.; Blonk, H.; Marinussen, M.; Van Zeist, W.; Starmans, D. *Methodology Used in Feedprint: A Tool Quantifying Greenhouse Gas Emissions of Feed Production and Utilization*; 1570-8616; Wageningen UR Livestock Research: Wageningen, The Netherlands, 2013.
22. Apajalahti, J.; Vienola, K.; Raatikainen, K.; Holder, V.; Moran, C.A. Conversion of branched-chain amino acids to corresponding isoacids-an in vitro tool for estimating ruminal protein degradability. *Front. Vet. Sci.* **2019**, *6*, 311. [CrossRef] [PubMed]
23. Meissner, H.; Henning, P.; Leeuw, K.-J.; Hagg, F.; Horn, C.; Kettunen, A.; Apajalahti, J. Efficacy and mode of action of selected non-ionophore antibiotics and direct-fed microbials in relation to Megasphaera elsdenii NCIMB 41125 during in vitro fermentation of an acidosis-causing substrate. *Livest. Sci.* **2014**, *162*, 115–125. [CrossRef]
24. Salami, S.A.; Moran, C.A.; Warren, H.E.; Taylor-Pickard, J. A Meta-Analysis of the Effects of Slow-Release Urea Supplementation on the Performance of Beef Cattle. *Animals* **2020**, *10*, 657. [CrossRef] [PubMed]
25. Apajalahti, J.H.; Kettunen, H.; Kettunen, A.; Holben, W.E.; Nurminen, P.H.; Rautonen, N.; Mutanen, M. Culture-independent microbial community analysis reveals that inulin in the diet primarily affects previously unknown bacteria in the mouse cecum. *Appl. Environ. Microbiol.* **2002**, *68*, 4986–4995. [CrossRef] [PubMed]
26. Weatherburn, M. Phenol-hypochlorite reaction for determination of ammonia. *Anal. Chem.* **1967**, *39*, 971–974. [CrossRef]
27. Berthelot, M. Violet d'aniline. *Rep. Chim. Appl.* **1859**, *1*, 284.
28. AOAC. *Association of Official Analytical Chemists*, 16th ed.; Official Methods of Analysis: Gaithersburg, MD, USA, 1996.
29. Ørskov, E.; McDonald, I. The estimation of protein degradability in the rumen from incubation measurements weighted according to rate of passage. *J. Agric. Sci.* **1979**, *92*, 499–503. [CrossRef]
30. SAS. *Stat User's Guide*; Cary, N.C., Ed.; Statistical Analysis Systems Inst. Inc.: Cary, NC, USA, 2001.
31. NRC. *Nutrient Requirements of Beef Cattle: Update 2000*; National Academies Press: Washington, DC, USA, 2000.
32. Van Soest, P.v.; Robertson, J.; Lewis, B. Methods for dietary fiber, neutral detergent fiber, and nonstarch polysaccharides in relation to animal nutrition. *J. Dairy Sci.* **1991**, *74*, 3583–3597. [CrossRef]
33. Lumsden, J.; Mullen, K.; Rowe, R. Hematology and biochemistry reference values for female Holstein cattle. *Can. J. Comp. Med.* **1980**, *44*, 24. [PubMed]
34. Martin, S.; Lumsden, J. The relationship of hematology and serum chemistry parameters to treatment for respiratory disease and weight gain in Ontario feedlot calves. *Can. J. Vet. Res.* **1987**, *51*, 499. [PubMed]
35. Mohri, M.; Sharifi, K.; Eidi, S. Hematology and serum biochemistry of Holstein dairy calves: Age related changes and comparison with blood composition in adults. *Res. Vet. Sci.* **2007**, *83*, 30–39. [CrossRef] [PubMed]
36. Salami, S.A.; Moran, C.A.; Warren, H.E.; Taylor-Pickard, J. Meta-analysis and sustainability of feeding slow-release urea in dairy production. *PLoS ONE* **2021**, *16*, e0246922. [CrossRef] [PubMed]
37. Kertz, A. Urea feeding to dairy cattle: A historical perspective and review. *Prof. Anim. Sci.* **2010**, *26*, 257–272. [CrossRef]
38. Loor, J.; Elolimy, A.; McCann, J. Dietary impacts on rumen microbiota in beef and dairy production. *Anim. Front.* **2016**, *6*, 22–29. [CrossRef]
39. Getachew, G.; Robinson, P.; DePeters, E.; Taylor, S. Relationships between chemical composition, dry matter degradation and in vitro gas production of several ruminant feeds. *Anim. Feed Sci. Technol.* **2004**, *111*, 57–71. [CrossRef]
40. Menke, K.H. Estimation of the energetic feed value obtained from chemical analysis and in vitro gas production using rumen fluid. *Anim. Res. Dev.* **1988**, *28*, 7–55.
41. Akay, V.; Tikofsky, J.; Holtz, C.; Dawson, K. Optigen®1200: Controlled release of non-protein nitrogen in the rumen. In Proceedings of the 20th Alltech Symposium, Lexington, KY, USA, 23–24 May 2004; Alltech Inc.: Nicholasville, KY, USA, 2004; pp. 179–185.
42. Bartley, E.; Davidovich, A.; Barr, G.; Griffel, G.; Dayton, A.; Deyoe, C.; Bechtle, R. Ammonia toxicity in cattle. I. Rumen and blood changes associated with toxicity and treatment methods. *J. Anim. Sci.* **1976**, *43*, 835–841. [CrossRef] [PubMed]
43. Chegeni, A.; Li, Y.; Deng, K.; Jiang, C.; Diao, Q. Effect of dietary polymer-coated urea and sodium bentonite on digestibility, rumen fermentation, and microbial protein yield in sheep fed high levels of corn stalk. *Livest. Sci.* **2013**, *157*, 141–150. [CrossRef]
44. Kertz, A.; Davidson, L.; Cords, B.; Puch, H. Ruminal infusion of ammonium chloride in lactating cows to determine effect of pH on ammonia trapping. *J. Dairy Sci.* **1983**, *66*, 2597–2601. [CrossRef]
45. Calsamiglia, S.; Blanch, M.; Ferret, A.; Moya, D. Is subacute ruminal acidosis a pH related problem? Causes and tools for its control. *Anim. Feed Sci. Technol.* **2012**, *172*, 42–50. [CrossRef]
46. Kenny, D.; Fitzsimons, C.; Waters, S.; McGee, M. Invited review: Improving feed efficiency of beef cattle–the current state of the art and future challenges. *Animal* **2018**, *12*, 1815–1826. [CrossRef] [PubMed]
47. McGrath, J.; Duval, S.M.; Tamassia, L.F.; Kindermann, M.; Stemmler, R.T.; de Gouvea, V.N.; Acedo, T.S.; Immig, I.; Williams, S.N.; Celi, P. Nutritional strategies in ruminants: A lifetime approach. *Res. Vet. Sci.* **2018**, *116*, 28–39. [CrossRef] [PubMed]
48. Bourg, B.; Tedeschi, L.; Wickersham, T.; Tricarico, J. Effects of a slow-release urea product on performance, carcass characteristics, and nitrogen balance of steers fed steam-flaked corn. *J. Anim. Sci.* **2012**, *90*, 3914–3923. [CrossRef] [PubMed]
49. Inostroza, J.; Shaver, R.; Cabrera, V.; Tricárico, J. Effect of diets containing a controlled-release urea product on milk yield, milk composition, and milk component yields in commercial Wisconsin dairy herds and economic implications. *Prof. Anim. Sci.* **2010**, *26*, 175–180. [CrossRef]
50. Radkowska, I.; Herbut, E. Hematological and biochemical blood parameters in dairy cows depending on the management system. *Anim. Sci. Papers Rep.* **2014**, *32*, 317–325.

51. Roland, L.; Drillich, M.; Iwersen, M. Hematology as a diagnostic tool in bovine medicine. *J. Vet. Diagn. Investig.* **2014**, *26*, 592–598. [CrossRef] [PubMed]
52. Lobley, G.; Connell, A.; Lomax, M.A.; Brown, D.S.; Milne, E.; Calder, A.G.; Farningham, D. Hepatic detoxification of ammonia in the ovine liver: Possible consequences for amino acid catabolism. *Br. J. Nutr.* **1995**, *73*, 667–685. [CrossRef] [PubMed]
53. Symonds, H.; Mather, D.L.; Collis, K. The maximum capacity of the liver of the adult dairy cow to metabolize ammonia. *Br. J. Nutr.* **1981**, *46*, 481–486. [CrossRef] [PubMed]
54. Word, J.; Martin, L.; Williams, D.; Williams, E.; Panciera, R.; Nelson, T.; Tillman, A. Urea toxicity studies in the bovine. *J. Anim. Sci.* **1969**, *29*, 786–791. [CrossRef]
55. Holder, V.B.; El-Kadi, S.W.; Tricarico, J.M.; Vanzant, E.S.; McLeod, K.R.; Harmon, D.L. The effects of crude protein concentration and slow release urea on nitrogen metabolism in Holstein steers. *Arch. Anim. Nutr.* **2013**, *67*, 93–103. [CrossRef]
56. Ribeiro, M.N.; Ribeiro, N.L.; Bozzi, R.; Costa, R.G. Physiological and biochemical blood variables of goats subjected to heat stress–a review. *J. Appl. Anim. Res.* **2018**, *46*, 1036–1041. [CrossRef]
57. Huntington, G.; Archibeque, S. Practical aspects of urea and ammonia metabolism in ruminants. *Proc. Am. Soc. Anim. Sci.* **2000**, *77*, 1–11. [CrossRef]

 *sustainability*

Review

Selenium: An Essential Micronutrient for Sustainable Dairy Cows Production

Hammad Ullah [1], Rifat Ullah Khan [2,*] , Vincenzo Tufarelli [3,*] and Vito Laudadio [3]

[1] College of Animal Husbandry & Veterinary Sciences, Abdul Wali Khan University, Mardan 23200, Pakistan; hammadullah@awkum.edu.pk

[2] Faculty of Animal Husbandry & Veterinary Sciences, College of Veterinary Sciences, The University of Agriculture, Peshawar 25130, Pakistan

[3] Department of DETO, Section of Veterinary Science and Animal Production, University of Bari 'Aldo Moro', Valenzano, 70010 Bari, Italy; vito.laudadio@uniba.it

* Correspondence: rukhan@aup.edu.pk (R.U.K.); vincenzo.tufarelli@uniba.it (V.T.)

Received: 24 November 2020; Accepted: 17 December 2020; Published: 21 December 2020

Abstract: This review article discusses the importance and effects of Selenium (Se) in sustainable dairy cows' production. The Se is an important micronutrient in dairy cows. It is one of the important feed derived antioxidants. It participates in important enzymes and enzyme reactions to improve metabolism, growth, and the defense system of the body, which results in the improved health of animals, particularly that of the mammary gland and reproductive system, thereby improving productive and reproductive performance. The Se is usually deficient in soil due to current extensive farming strategies, so its supplementation is generally advised. Supplementation of Se in organic form is generally preferred over inorganic form due to its better incorporation and uptake, resulting in improved performance. Kidneys, liver, testis, and lungs are important sites for Se storage. The Se is excreted in urine, feces, exhaled breath, loss of skin, and hair cells. Although Se supplementation plays an important role in the profitability of dairy cows, its excess intake is toxic and should be avoided.

Keywords: selenium; cow; production; reproduction; nutrition

---

1. Introduction

Selenium (Se) is a nonmetal with the atomic number 34. In the periodic table, the element is located in the fourth period [1,2]. Se was discovered by a Swedish chemist, Jons Jacob Berzelius, in the year 1817 [3]. It is regarded as trace element due to its low content (0.05 ppm) in the earth's crust [4,5]. The Se can occur in organic as well as inorganic form [6]. Inorganic forms of Se include metal selenides, elemental selenium, and selenates (IV) and (VI). Organic forms of Se include selenium amino acids, methyl compounds, selenoproteins, selenocysteine, and selenomethionine [7].

Se was first reported as an essential nutrient for animals by Schwarz and Foltz [8]. In biological samples, Se is present in the form of selenomethionine and selenocysteine [9]. Se plays an important role in animals and human health [10]. In the human body, Se plays a significant role in various biochemical and physiological processes [11].

In farm animals, adequate Se intake prevents various disorders like white muscle disease, mulberry heart disease, dietary necrotic liver degeneration, parturition problems, retention of placenta, *post*-parturient paraplegia, and early embryonic mortality [12]. In cattle, Se supplementation reduces postpartum reproductive disorders like ovarian cysts and metritis [13]. Se also plays an important role in udder health, preventing clinical and subclinical mastitis [14]. Se is a structural component of important proteins involved in defense mechanisms [15]. Further, Se occupies an important part in these proteins [16]. The body cannot produce Se by itself. Forage is a natural source of Se for animals [17]. Selenium is usually deficient in the soil, and the current extensive farming strategies tend to favor Se deficiency [18]. The soil Se concentration differs to a great extent even within small areas.

Therefore, in livestock animals, Se supplementation is advised so that the minimum intake level is certain [19]. Dairy cattle require Se at the rate of 300 μg/kg DM [13]. Recently, several studies have recommended that organic Se is superior to the inorganic form of Se in dairy cattle [20].

2. Absorption, Distribution, Metabolism, and Excretion of Se

Mechanisms of Se homeostasis are important because the element is potentially toxic as well as an essential micronutrient [21]. The Se is absorbed from the small intestine. Organic Se is obtained by the body from selenized yeast and basal feed ingredients in the form of seleno-amino acids (i.e., selenocysteine and selenomethionine), while inorganic selenium supplementation provides selenate and selenite. Seleno-methionine (Se-Methionine) is absorbed via methionine transporter system, selenate is absorbed by active transport system while absorption of selenite mostly takes place by passive diffusion [22]. Rate of absorption of organic Se is higher than inorganic Se [23]. Inorganic Se is mostly reduced to insoluble elemental Se or is readily absorbed into feed particles in the digestive tract [24], resulting in excretion of most of the Se [25]. In contrast, most of organic Se (in form of Se-Methionine) is incorporated in rumen microorganisms so less elemental Se is formed.

Oral bioavailability of Se-Methionine is, therefore, greater than that of inorganic Se [26]. In blood, Se is bound to low density lipoprotein (LDL), very low-density lipoprotein (VLDL), albumin, and α and β-globulins [27]. Most of the Se in body fluids and tissues is present either in the form of Se-methionine or selenocysteine [28]. In cattle, kidneys are the site of the highest density of Se, and muscles are the site of the highest concentration of Se [29]. The highest amount of Se is stored in the kidneys followed by the liver, testis, and lungs [30]. Liver, heart, and skeletal muscles are most sensitive to Se deficiency [29]. The liver is considered the Se storage organ to which a considerable amount of Se is directed during absorption for accumulation [31]. In cattle, concentration of Se rises in serum two to six days following increased supplementation in diet [32].

Various organic and inorganic Se sources are first transformed to inorganic selenide before synthesis of Se-cysteine which, in turn, contributes to the bioactive component of Selenoproteins. Following absorption, Se-Methionine can be found in plasma methionine pool and blood proteins as it is transported to body tissues. One of example of the transportation is extraction of large amounts of methionine by the mammary gland for milk protein synthesis, resulting in a large quantity of Se in milk, which is beneficial to the neonate and human consumers [33]. With an increase in Se intake in Se deficient animals, the concentration of Se in tissues and the whole body rises sharply due to accumulation of Se in tissues as selenoproteins until adequate body Se status is achieved. However, the rate of Se excretion increases once the body requirement for Se has been fulfilled, thereby reducing its further accumulation [34]. The liver plays an important role in Se excretion [35]. Excretion of Se is important for regulation of whole-body Se [34]. Se is excreted in urine, feces, exhaled breath, loss of skin, and hair cells [36]. Polymethylation of selenide takes place before excretion. Dimethyl selenide is excreted in feces and breath, while the cation $(CH_3)_3Se^+$ is excreted in urine [31]. With an increase in the amount of dietary Se supplementation, the concentration of Se excretion in urine and feces increases. However, the form of Se intake does not influence the concentration of Se excreted [37].

3. Role of Se in Antioxidant Defense Mechanism

Oxidation is a process in which loss of electrons occurs [38]. Oxidation reactions are very important for life, but they can also have detrimental effects [39]. These reactions may produce reactive oxygen species (ROS) [40]. ROS can cause oxidative damage and cell death in case of excessive production [41]. ROS damage macromolecules of the cells leading to lipid peroxidation, nucleic acid, and protein alterations. Formation of ROS is considered as a pathobiochemical mechanism involved in progression or initiation phase of various diseases. To maintain correct cell signaling, radical scavenging enzymes must maintain a threshold level ROS inside the cell. Otherwise, increased ROS production not only causes excessive signals to the cells, but also directly damages key components in signaling pathways [42]. Animals with high productive performance are more susceptible to oxidative stress, resulting in their lower productive and reproductive performance [43].

Under stress conditions, the antioxidant system of the body requires dietary antioxidant supplementation via water/feed, as it cannot deal with excessive ROS formation properly [44]. Antioxidant is a natural or man-made substance that may prevent or delay some types of cell damage [39]. Antioxidants play a significant role in the body's defense system against reactive oxygen species (ROS) [45]. Se is one of the important feed derived antioxidants [43]. Se is an important part of 25 selenoproteins identified in animals [43]. More than half of these proteins are involved in redox balance and antioxidant defense of the body [44]. Important families of selenoproteins include iodothyronine deiodinases, thioredoxin reductases, and glutathione peroxidases (GPX) [45]. The GPX, for example, protects cells against oxidative injury [46]. Activity of this enzyme is better in cows fed Se supplemented diets compared to cows fed Se deficient diets [47–49].

To kill phagocytized bacteria, neutrophils must provide a high oxidizing intracellular environment, but at the same time they must also maintain a balance between reactive oxygen metabolites (H_2O_2 and superoxide [O^{2-}]) so that cell damage and death can be avoided [33]. In neutrophils of Se deficient cattle, GPX activity and oxygen consumption is lower than normal [50]. This results in oxidative injury of body cells [46]. Due to healthy GPX activity, destruction of cellular proteins and necrosis can be avoided, as it neutralizes the effects of lipid hydroperoxide and hydrogen peroxide [51]. Organic Se is superior to inorganic Se in improving antioxidant status in dairy animals [44].

4. Role of Se in Milk Production

General health of the animal, health of the mammary gland, environmental conditions such as insufficient nutrient intake, intensity of production system, and low corporal condition directly influence milk production [52]. Production can be improved in lactating dairy cows by supplementation of organic trace minerals [53]. Se is one of the trace minerals [54]. It is an essential element in ruminant nutrition [54,55]. Deficiency of Se impairs the immune function, leading to a higher risk of illness including that of the udder, which results in lower milk production [56].

Se helps to decrease linear somatic cell count scores [14]. Se affects innate and adaptive immune responses of the mammary gland through humoral and cellular activities [57]. It improves the bactericidal effects of milk and blood neutrophils [58]. It, therefore, prevents the incidence of mastitis [59]. In short, higher milk production in response to supplementation of Se is due to improvement in immune system brought about by Se [60].

Se yeast also improves feed digestibility, resulting in improved milk production [61]. There is a better incorporation and uptake of selenium when it is supplemented as selenium yeast [56]. Yeast based Se retains higher in tissues than inorganic Se and ensures sufficient Se availability for disease reduction [62]. Supplementation of Se in yeast form increases milk production by 24.8% [63]. When Se is supplemented along with vit E, a payback of 0.21 US cents/animal/day can be increased [64].

5. Role of Se in Feed Intake, Feed Utilization, and Body Growth

Feed intake is a critical factor for milk production of dairy cattle. The intake generally declines under heat stress to reduce metabolic heat production [65]. Feeding management is one of the key factors to enhance feed intake of dairy cows. [66]. Supplementation of diet with supranutritional Se and vit E enhances feed intake in heat stress condition by improving antioxidant status and thyroid hormone activity [67]. Se supplementation also improves growth performance in heifers, especially during the early stages of growth [68]. Se participates in important enzymes and enzyme reactions [69,70]. For example, Se helps in the expression of iodothyronine deiodinases, which regulate active T3 hormone production in peripheral tissues and the thyroid gland to improve metabolism and growth [71]. Thyroid peroxidase is also a selenoenzyme important for the process of iodization of globulin, thus avoiding damage to the thyroid epithelial cell membrane [72,73].

Supplementing Se enhances organic matter (OM) and crude protein (CP) intake [74]. Se supplementation improves total digestible nutrients (TDN) intake [68]. A higher TDN intake improves microbial efficiency in cattle. Due to differences in metabolism, the organic form of Se has better bioavailability compared to the inorganic form [75]. Supplementation of Se in yeast form improves digestibility of dry matter (DM), organic matter (OM), crude protein, neutral detergent fiber (NDF), and acid detergent fiber [76]. Se yeast improves feed digestibility in lactating dairy cows [61].

6. Role of Se in Cattle Reproductive Performance

Reproductive performance, such as age at first calving and calving interval, are important for improving productivity and profitability of dairy cows [77]. Trace minerals play very important roles in improving reproductive health and performance [78]. Fertility in dairy cows is getting worse in recent years, so enhancement of the conception rate has become an important issue. Although multiple factors are responsible for the declining fertility, improving the function of corpus luteum (CL) is one of the focuses of recent studies [79]. The function of CL is very important for reproduction according to many researchers [79]. For example, it was found that a reduced level of progesterone (P4) in plasma during the ovulatory follicles' growth is associated with decreased conception rate [80], and a low level of plasma progesterone in the blood was associated with low survivability of the embryo during early pregnancy [81]. Progesterone concentration in blood plasma of dairy cows supplemented with selenium enriched yeast shows an early increase compared to cows to which no selenium is fed [79]. Se increases proliferation of bovine corpus luteal cells by degrading lipid peroxides resulting in increased progesterone concentration [82]. This improved luteal function helps in preventing early embryonic death [83]. So, Se supplementation improves fertility by reducing embryonic death during the first month of gestation [13].

Se supplementation also improves estrous percentage and results in lower age at first conception [84]. Supplementation of diet with organic Se in lactation and transition period improves the immune function and health of the uterus followed by improved reproductive performance [85]. Pregnancy enhances oxidative stress and vitamin E, and Se supplementation reduces the stress [86]. Deficiency of Se is usually associated with a high risk of metritis, retention of placenta, abortion, increased susceptibility to infections, and lower fertility [87]. Se supplementation improves the rate of reproduction. However, results are best when Se is supplemented as Se yeast. There is a better incorporation and uptake of selenium when it is supplemented as Selenium yeast [56].

7. Toxicity and Concluding Remarks

Se is an indispensable micronutrient in animals and is a fundamental part of seleno-enzymes. However, its excess is toxic for animals [88]. Toxicity of Se was described long before its actual discovery. In 1295, Marco Polo observed that in the north-eastern region of China, hooves of animals consuming a particular type of plant became deformed [89]. At the turn of the nineteenth century, it was discovered by US researchers that similar situation arose in cattle due to eating plants containing Se in large quantities [10,90].

Different factors, like routes of administration of Se, species of animals, and chemical forms of the element, influence the severity of Se poisoning. The soluble salts of Se (Na_2SeO_3 and $Na2SeO_4$) are among the more toxic compounds; seleno-aminoacids and Se inherent in grains are relatively moderately toxic; the poorly soluble forms (e.g., diphenyl selenide, SeS_2, Na_2Se, and elemental Se) are among the least toxic of the Se compounds. Oral administration of Se poses less chances of toxicity than parental administration [91].

Chronic selenosis, often called alkali disease and acute selenosis, popularly known as blind staggers, are the most common form of selenosis [12]. Animals consuming feed containing Se at the rate of 5–8 mg/kg of dry matter are at risk of Se toxicity [92,93]. Livestock within 1 km of toxic farms is at risk of the toxicity. Therefore, blood samples from the livestock within 5 miles of known toxic farms should be tested for levels of Se before supplementation with the mineral [94].

Acute poisoning is not common due to the unpalatability of plants with high Se levels. However, hungry animals may eat such plants and thus suffer from acute poisoning. In cattle, acute poisoning is characterized by watery diarrhea, abnormal posture and movement, labored respiration, indications of abdominal pain, elevated temperature, prostration, and death. Chronic selenium poisoning may result from consumption of seleniferous feed over prolonged periods such as weeks or months. Important signs of the poisoning in cattle include hoof malformations, emaciation, loss of hair, and lameness [95]. In advanced cases, atrophy of the heart, anemia, and liver cirrhosis occur [91]. Excess Se intake impairs some immunological functions [96].

It is concluded from the present review that Se supplementation, especially in organic form, improves immunity, feed utilization, reproductive performance, and milk production in dairy cattle (Figure 1).

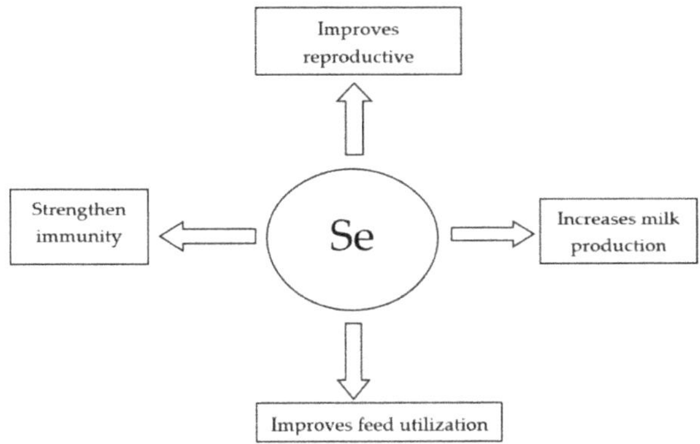

Figure 1. Benefits of Se supplementation in dairy cattle nutrition.

Lastly, excessive antioxidant supplementation generally can lead to an increased production of ROS [97,98]. Inorganic Se (selenite) is more toxic than the organic form. Inorganic Se reacts with glutathione, resulting in formation of selenotrisulphides which, in turn, react with other thiols and thus lead to the generation of oxygen free radicals, for example, superoxide anion. Organic Se is converted into selenols. This also results in the generation of oxygen free radicals inducing oxidative stress [88]. Mineral sulphates given in high doses can antagonize Se [94]. Deficiency of Se [99] is a much more common problem than its toxicity.

Table 1 summarizes the effects of the different forms of Se supplementation on productive and reproductive parameters, as well as the health status of dairy cows.

Table 1. Effect of Se supplementation on different parameters of dairy cows. DM: Dry matter; OM: Organic matter; NDF: Neutral detergent fiber; NFC: Non-fiber carbohydrates.

Parameter	Form of Se	Dose of Se	Effect	Reference
Milk production	Organic	0.15 mg/kg DMI	Increased	[61]
	Organic	0.3 mg/kg DMI	Increased	[61,63,64]
	Organic	0.4 mg/kg DMI	Decreased	[63]
Neutrophil function	Inorganic/sodium selenite	10 μM	Improved	[58]
	Organic	0.3 mg/kg DMI	Increased	[33]
Adaptive immunity	Organic	0.3 mg/kg DMI	Improved	[33]
Somatic cell count	Organic	0.3 mg/kg DMI	Decreased	[33]

Table 1. *Cont.*

Parameter	Form of Se	Dose of Se	Effect	Reference
Milk fat	Organic	0.3 mg/kg DMI	Increased	[100]
	Nano particles	0.3 mg/kg DMI	Increased	[100]
	Organic	0.3 mg/kg DMI	Increased	[33]
Milk protein	Organic	0.3 mg/kg DMI	Increased	[100]
	Nano particles	0.3 mg/kg DMI	Increased	[100]
Digestibility (DM, OM, NDF, Ether extract, NFC)	Organic	0.3 mg/kg DMI	Increased	[100]
	Nano particles	0.3 mg/kg DMI	Increased	[100]
Plasma P4	Inorganic	0.5 mg/kg DMI	Increased	[101]
Serum Se	Organic + Inorganic	105 mg/week + 0.3 mg/kg DMI	Increased	[102]
Plasma Se	Organic	0.3 mg/kg DMI	Increased	[33]
Whole blood Se concentration	Organic + Inorganic	105 mg/week + 0.3 mg/kg DMI	Increased	[102]
Serum amyloid A	Organic + Inorganic	105 mg/week + 0.3 mg/kg DMI	Increased	[102]
Erythrocyte glutathione	Organic + Inorganic	105 mg/week + 0.3 mg/kg DMI	Increased	[102]
Serum albumin concentration	Organic + Inorganic	105 mg/week + 0.3 mg/kg DMI	Increased	[102]
Serum cholesterol concentration	Organic + Inorganic	105 mg/week + 0.3 mg/kg DMI	Decreased	[102]
α-tocopherol/cholesterol ratio	Organic + Inorganic	105 mg/week + 0.3 mg/kg DMI	Increased	[102]
Uterine health	Organic	0.3 mg/kg DMI	Improved	[33]
Second service pregnancy rate	Organic	0.3 mg/kg DMI	increased	[33]

Author Contributions: All the authors equally contributed and commented on early and final versions of manuscript. All authors have read and agreed to the published version of the manuscript.

Funding: This research received no external funding.

Acknowledgments: Authors extend thanks to their respected institutes and universities.

Conflicts of Interest: The authors declare no conflict of interest.

References

1. Butterman, W.C.; Brown, R.D. *Mineral Commodity Profiles—Selenium*; U.S. USGS Numbered Series; USGS Publisher: Reston, Virginia, 2004; p. 20.
2. Perrone, D.; Monteiro, M.; Nunes, J.C. The Chemistry of Selenium. *R. Soc. Chem.* **2015**, 3–15. [CrossRef]
3. Akl, M.A.; Ismael, D.S.; El-Asmy, A.A. Precipitate flotation-separation, speciation and hydride generation atomic absorption spectrometric determination of selenium (IV) in food stuffs. *Microchem. J.* **2006**, *83*, 61–69. [CrossRef]
4. Shrivas, K.; Patel, D.K. Ultrasound assisted-hollow fibre liquid-phase microextraction for the determination of selenium in vegetable and fruit samples by using GF-AAS. *Food Chem.* **2011**, *124*, 1673–1677. [CrossRef]
5. Fan, A.M.; Kizer, K.W. Selenium-nutritional, toxicologic, and clinical aspects. *West. J. Med.* **1990**, *153*, 160–167. [PubMed]
6. Vonderheide, A.P.; Wrobel, K.; Kannamkumarath, S.S.; B'Hymer, C.; Montez-Bay'on, M.; Ponce de Leon, C.; Caruso, J.A. Characterization of selenium species in Brazil nuts by HPLC-ICP-MS and ES-MS. *J. Agric. Food Chem.* **2002**, *50*, 5722–5728. [CrossRef] [PubMed]

7. Najafi, N.M.; Seidi, S.; Alizadeh, R.; Tavakoli, H. Inorganic selenium speciation in environmental samples using selective electrodeposition coupled with electrothermal atomic absorption spectrometry. *Spectrochim. Acta B* **2010**, *65*, 334–339. [CrossRef]

8. Schwarz, K.; Foltz, C.M. Selenium as an integral part of factor-3 against dietary necrotic liver degeneration. *J. Am. Chem. Soc.* **1957**, *79*, 3292. [CrossRef]

9. Gergely, V.; Montes-Bayon, M.; Fodor, P.; Sanz-Medel, A. Selenium species in aqueous extracts of alfalfa sprouts by two-dimensional liquid chromatography coupled to inductively coupled plasma mass spectrometry and electrospray mass spectrometry detection. *J. Agric. Food Chem.* **2006**, *54*, 4524–4530. [CrossRef]

10. Bodnar, M.; Konieczka, P.; Namieśnik, J. The Properties, Functions, and Use of Selenium Compounds in Living Organisms. *J. Environ. Sci. Health Part C Environ. Carcinog. Ecotoxicol. Rev.* **2012**, *30*, 225–252. [CrossRef]

11. Puzanowska-Tarasiewicz, H.; Kuzmicka, L.; Tarasiewicz, M. Biological functions of selected elements and their compounds. 3, Zinc—An enzyme component and activator. *Pol. Med. Merkur. Organ Pol. Med. Soc.* **2009**, *27*, 419–422.

12. Zarczy'nska, K.; Sobiech, P.; Radwi 'nska, J.; Rękawek, W. Effects of selenium on animal health. *J. Elementol.* **2013**, *18*, 329–340.

13. Mehdi, Y.; Dufrasne, I. Selenium in Cattle: A Review. *Molecules* **2016**, *21*, 545. [CrossRef] [PubMed]

14. Machado, V.S.; Bicalho, M.L.; Pereira, R.V.; Caixeta, L.S.; Knauer, W.A.; Oikonomou, G.; Gilbert, R.O.; Bicalho, R.C. Effect of an injectable trace mineral supplement containing selenium, copper, zinc, and manganese on the health and production of lactating Holstein cows. *Vet. J.* **2013**, *197*, 451–456. [CrossRef] [PubMed]

15. Brigelius-Flohé, R.; Maiorino, M. Glutathione peroxidases. *Biochim. Biophys. Acta* **2013**, *1830*, 3289–3303. [CrossRef] [PubMed]

16. Mehdi, Y.; Hornick, J.L.; Istasse, L.; Dufrasne, I. Selenium in the environment, metabolism and involvement in body functions. *Molecules* **2013**, *18*, 3292–3311. [CrossRef] [PubMed]

17. López-Alonso, M. Trace minerals and livestock: Not too much not too little. *Int. Sch. Res. Netw. Vet. Sci.* **2012**, 704825. [CrossRef] [PubMed]

18. Schöne, F.; Steinhöfel, O.; Weigel, K.; Bergmann, H.; Herzog, E.; Dunkel, S.; Kirmse, R.; Leiterer, M. Selenium in feedstuffs and rations for dairy cows including a view of the food chain up to the consumer. *J. Verbrauch. Lebensm.* **2013**, *8*, 271–280.

19. Lu, J.; Holmgren, A. Selenoproteins. *J. Biol. Chem.* **2009**, *284*, 723–727. [CrossRef]

20. Slavik, P.; Illek, J.; Brix, M.; Hlavicoca, J.; Rajmon, R.; Jilek, F. Influence of organic versus inorganic dietary selenium supplementation on the concentration of selenium in colostrum, milk and blood of beef cows. *Acta Vet. Scand.* **2008**, *50*, 1–6. [CrossRef]

21. Thomson, C.D.; Stewart, R.D. Metabolic studies of (75Se) selenomethionine and (75Se) selenite in the rat. *Br. J. Nutr.* **1973**, *30*, 139–147. [CrossRef]

22. Weiss, W.P. Selenium nutrition of dairy cows: Comparing responses to organic and inorganic selenium forms. In *Nutritional Biotechnology in the Feed and Food Industries, Proceedings of Alltech's*; Lyons, P.T., Jacques, K.A., Eds.; Nottingham University Press: Nottingham, UK, 12–14 May 2003; pp. 333–343.

23. Tufarelli, V.; Ceci, E.; Laudadio, V. 2-Hydroxy-4-methylselenobutanoic acid as new organic selenium dietary supplement to produce selenium-enriched eggs. *Biol. Trace Elem. Res.* **2016**, *171*, 453–458. [CrossRef] [PubMed]

24. Podoll, K.L.; Bernard, J.B.; Ullrey, D.E.; DeBar, S.R.; Ku, P.K.; Magee, W.T. Dietary selenate versus selenite for cattle, sheep, and horses. *J. Anim. Sci.* **1992**, *70*, 1965–1970. [CrossRef] [PubMed]

25. Lopez, P.L.; Preston, R.L.; Pfander, W.H. Wholebody retention, tissue distribution and excretion of selenium after oral and intravenous administration in lambs fed varying selenium intakes. *J. Nutr.* **1969**, *97*, 123–132. [CrossRef] [PubMed]

26. Galbraith, M.L.; Vorachek, W.R.; Estill, C.T.; Whanger, P.D.; Bobe, G.; Davis, T.Z.; Hall, J.A. Rumen microorganisms decrease bioavailability of inorganic selenium supplements. *Biol. Trace Elem. Res.* **2015**, *171*, 338–343. [CrossRef]

27. Schrauzer, G.N. Selenomethionine: A review of its nutritional significance, metabolism and toxicity. *J. Nutr.* **2000**, *130*, 1653–1656. [CrossRef]

28. Suzuki, K.T.; Ogra, Y. Metabolic pathway for selenium in the body: Speciation by HPLC-ICP MS with enriched Se. *Food Addit. Contam.* **2002**, *19*, 974–983. [CrossRef]
29. Meschy, F. *Nutrition Minérale des Ruminants*; Editions Quae: Versaille, France, 2010; p. 208.
30. Juniper, D.T.; Phipps, R.H.; Ramos-Morales, E.; Bertin, G. Effect of dietary supplementation with selenium-enriched yeast or sodium selenite on selenium tissue distribution and meat quality in beef cattle. *J. Anim. Sci.* **2008**, *86*, 3100–3109. [CrossRef]
31. Combs, G.F. Biomarkers of selenium status. *Nutrients* **2015**, *7*, 2209–2236. [CrossRef]
32. Ellis, R.G.; Herdt, T.H.; Stowe, H.D. Physical, hematologic, biochemical, and immunologic effects of supranutritional supplementation with dietary selenium in Holstein cows. *Am. J. Vet. Res.* **1997**, *58*, 760–764.
33. Silvestre, F.T.; Rutigliano, H.M.; Thatcher, W.W.; Santos, J.E.P.; Staples, C.R. *Florida Ruminant Nutrition Symposium*; University of Florida: Gainesville, FL, USA, 2007; pp. 48–61.
34. Lucia, F.; Pedrosa, C.; Motley, A.K.; Stevenson. T.D.; Hill, K.E. Fecal Selenium Excretion Is Regulated by Dietary Selenium Intake. *Biol. Trace Elem. Res.* **2012**, *149*, 377–381.
35. Burk, R.F.; Hill, K.E. Regulation of Selenium Metabolism and Transport. *Ann. Rev. Nutr.* **2015**, *35*, 109–134. [CrossRef] [PubMed]
36. Hopkins, L.L., Jr.; Pope, A.L.; Baumann, C.A. Distribution of microgram quantities of selenium in the tissues of the rat, effects of previous selenium intake. *J. Nutr.* **1966**, *88*, 61–65. [CrossRef] [PubMed]
37. Juniper, D.T.; Phipps, R.H.; Jones, A.K.; Bertin, G. Selenium Supplementation of Lactating Dairy Cows: Effect on Selenium Concentration in Blood, Milk, Urine, and Feces. *J. Dairy Sci.* **2006**, *89*, 3544–3551. [CrossRef]
38. Hrbac, J.; Kohen, R. Biological redox activity: Its importance, methods for its quantification and implication for health and disease. *Drug Dev. Res.* **2000**, *50*, 516–527. [CrossRef]
39. Yadav, A.; Kumari, R.; Yadav, A.; Mishra, J.P.; Srivatva, S.; Prabha, S. Antioxidants and its functions in human body—A Review. *Res. Environ. Life Sci.* **2016**, *9*, 1328–1331.
40. Hancock, J.T.; Desikan, R.; Neill, S.J. Role of Reactive Oxygen Species in Cell Signaling Pathways. *Biochem. Biomed. Asp. Oxidative Modif.* **2001**, *29*, 345–350.
41. Sharma, P.; Jha, A.B.; Dubey, R.S.; Pessarakli, M. Reactive Oxygen Species, Oxidative Damage, and Antioxidative Defense Mechanism in Plants under Stressful Conditions. *J. Bot.* **2012**, 217037. [CrossRef]
42. Adwas, A.A.; Elsayed, A.; Azab, A.E.; Quwaydir, F.A. Oxidative stress and antioxidant mechanisms in human body. *J. Appl. Biotechnol.* **2019**, *6*, 43–47.
43. Surai, P.F.; Kochish, I.I.; Fisinin, V.I.; Juniper, D.T. Revisiting Oxidative Stress and the Use of Organic Selenium in Dairy Cow Nutrition. *Animals* **2019**, *9*, 462. [CrossRef]
44. Surai, P.F. *Selenium in Poultry Nutrition and Health*; Academic Publishers: Wageningen, The Netherlands, 2018.
45. Boxin, O.U.; Dejian, H.; Maureen, A.F.; Elizabeth, K.D. Analysis of antioxidant activities of common vegetables employing oxygen radical Absorbance Capacity (ORAC) and Ferric Reducing Antioxidant Power (FRAP) Assays: A comparative study. *J. Agric. Food Chem.* **2002**, *5*, 223–228.
46. Burk, R.F.; Hill, K.E.; Motley, A.K. Selenoprotein Metabolism and Function: Evidence for More than One Function for Selenoprotein. *Pak. J. Nutr.* **2003**, *133*, 1517S–1520S. [CrossRef] [PubMed]
47. Mezzetti, A.; Ilio, C.D.; Calafiore, A.M.; Aceto, A.; Marzio, L.; Frederici, G.; Cuccurullo, F. Glutathione peroxidase, glutathione reductase and glutathione transferase activities in the human artery, vein and heart. *J. Mol. Cardiol.* **1990**, *22*, 935–938. [CrossRef]
48. Grasso, P.J.; Scholz, R.W.; Erskine, R.J.; Eberhart, R.J. Phagocytosis, bactericidal activity, and oxidative metabolism of mammary neutrophils from dairy cows fed selenium-adequate and selenium-deficient diets. *Am. J. Vet. Res.* **1990**, *51*, 269–274. [PubMed]
49. Gunter, S.A.; Beck, P.A.; Phillips, J.M. Effects of supplementary selenium source on the performance and blood measurements in beef cows and their calves. *J. Anim. Sci.* **2003**, *81*, 856–864. [CrossRef] [PubMed]
50. Arthur, J.R.; Boyne, R. Superoxide dismutase and glutathione peroxidase activities in neutrophils from selenium deficient and copper deficient cattle. *Life Sci.* **1985**, *36*, 1569–1575. [CrossRef]
51. Awadeh, F.T.; Kincaid, R.L.; Johnson, K.A. Effect of level and sources of dietary selenium on concentrations of thyroid hormones and immunoglobulins in beef cows and calves. *J. Anim. Sci.* **1998**, *76*, 1204–1215. [CrossRef] [PubMed]
52. Butler, W.R. Nutritional interactions with reproductive performance in dairy cattle. *Anim. Reprod. Sci.* **2000**, *60–61*, 449–457. [CrossRef]

53. Rabiee, A.R.; Lean, I.J.; Stevenson, M.A.; Socha, M.T. Effects of feeding organic trace minerals on milk production and reproductive performance in lactating dairy cows: A meta-analysis. *J. Dairy Sci.* **2010**, *93*, 4239–4251. [CrossRef]

54. Oltramari, C.E.; Pinheiro, M.G.; Miranda, M.S.; Arcaro, J.R.P.; Castelani, L.; Toledo, L.C.; Ambrósio, L.A.; Leme, P.R.; Manella, M.Q.; Júnior, I.A. Selenium Sources in the Diet of Dairy Cows and Their Effects on Milk Production and Quality, on Udder Health and on Physiological Indicators of Heat Stress. *Ital. J. Anim. Sci.* **2014**, *13*, 2921. [CrossRef]

55. Stevens, J.B. Serum selenium concentrations and glutathione peroxidase activities in cattle grazing forages of various selenium concentrations. *Am. J. Vet. Res.* **1985**, *46*, 1556–1561.

56. Tufarelli, V.; Laudadio, V. Dietary supplementation with selenium and vitamin E improves milk yield, composition and rheological properties of dairy Jonica goats. *J. Dairy Res.* **2011**, *78*, 144–148. [CrossRef] [PubMed]

57. Salman, S.; Khol-Parisini, A.; Schafft, H.; Lahrssen-Wiederholt, M.; Hulan, H.W.; Dinse, D.; Zentek, J. The role of dietary selenium in bovine mammary gland health and immune function. *Anim. Health Res. Rev.* **2009**, *10*, 21–34. [CrossRef] [PubMed]

58. Souza, F.N.; Blagitz, M.G.; Latorre, A.O.; Mori, C.S.; Sucupira, M.C.A.; Libera, A.M.M.P.D. Effects of in vitro selenium supplementation on blood and milk neutrophils from dairy cows. *Pesqui. Vet. Bras.* **2012**, *32*, 174–178. [CrossRef]

59. Khalifa, H.H.; Safwat, M.A.; El-Sysy, M.A.I.; Al-Metwaly, M.A. Effect of selenium and vitamin E supplementation as a nutritional treatment for some physiological and productive traits of Holstein dairy cows under Egyptian summer conditions. *J. Egypt. Acad. Soc. Environ. Dev.* **2016**, *17*, 97–113. [CrossRef]

60. Cordova-Izquierdo, C.G.R.L.; Campos, V.M.X.; Suarez, S.C.; Cordova-Jimenez, C.A.; Cordova-Jimenez, M.S. Effect of gonadotophin releasing factor and antioxidants on the rate of estrous repetition dairy cows. *Aust. J. Basic Appl. Sci.* **2010**, *408*, 1282–1287.

61. Wang, C.; Liu, Q.; Yang, W.Z.; Dong, Q.; Yang, X.M.; He, D.C.; Huang, Y.X. Effects of selenium yeast on rumen fermentation, lactation performance and feed digestibilities in lactating dairy cows. *Livest. Sci.* **2009**, *126*, 239–244. [CrossRef]

62. Wilde, D. Influence of macro and micro minerals in the peri-parturient period on fertility in dairy cattle. *Anim. Reprod. Sci.* **2006**, *96*, 240–249. [CrossRef]

63. Ullah, H.; Khan, R.U.; Mobashar, M.; Ahmad, S.; Sajid, A.; Khan, N.U.; Usman, T.; Khattak, I.; Hamayun, K. Effect of yeast-based selenium on blood progesterone, metabolites and milk yield in Achai dairy cows. *Ital. J. Anim. Sci.* **2019**, *18*, 1445–1450. [CrossRef]

64. Eulogio, G.L.J.; Alberto, S.O.J.; Hugo, C.V.; Antonio, C.N.; Alejandro, C.; Juan, M.Q. Effects of Selenium and vitamin E in the production, physiochemical composition and somatic cell count in milk of Ayrshire cows. *J. Anim. Vet. Adv.* **2012**, *11*, 687–691.

65. Chauhana, S.S.; Ponnampalama, E.N.; Celi, P.; Hopkins, D.L.; Leurya, B.J.; Dunsheaa, F.R. High dietary vitamin E and selenium improves feed intake and weight gain of finisher lambs and maintains redox homeostasis under hot conditions. *Small Rumin. Res.* **2016**, *137*, 17–23. [CrossRef]

66. Smith, J.F.; Brouk, M.J. Factors affecting dry matter intake by lactating dairy cows. *Kans. Agric. Exp. Stn. Res. Rep.* **2000**, *2*, 52–58. [CrossRef]

67. Alhidary, I.; Shini, S.; Al Jassim, R.; Gaughan, J. Effect of various doses of injected selenium on performance and physiological responses of sheep to heat load. *J. Anim. Sci.* **2012**, *90*, 2988–2994. [CrossRef] [PubMed]

68. Ganie, A.A.; Baghel, R.P.S.; Mudgal, V.; Sheikh, G.G. Effect of Selenium Supplementation on Growth and Nutrient Utilization in Buffalo Heifers. *Anim. Nutr. Feed Technol.* **2010**, *10*, 255–259.

69. Surai, P.F. Selenium in poultry nutrition-1. Antioxidant properties, deficiency and toxicity. *World Poult. Sci. J.* **2002**, *58*, 333–347. [CrossRef]

70. Willshire, J.A.; Payne, J.H. Selenium and vitamin E in dairy cows—A review. *Cattle Pract.* **2011**, *19*, 22–30.

71. Rooke, J.A.; Robinson, J.J.; Arthur, J.R. Effects of vitamin E and selenium on the performance and immune status of ewes and lambs. *J. Agric. Sci.* **2004**, *142*, 253–262. [CrossRef]

72. Gartner, R.; Gasnier, B.; Dietrich, J.; Krebs, B.; Angstwurm, M. Selenium supplementation in patients with autoimmune thyroiditis decreases thyroid peroxidase antibodies concentrations. *J. Clin. Endocrinol. Metab.* **2007**, *87*, 1687–1691. [CrossRef]

73. Schomburg, L.; Riese, C.; Michaelis, M.; Griebert, E.; Klein, M.; Sapin, R.; Schweizer, U.; Köhrle, J. Synthesis and metabolism of thyroid hormones is preferentially maintained in seleniumdeficient transgenic mice. *Endocrinology* **2007**, *147*, 1306–1313. [CrossRef]

74. Taheri, Z.; Karimi, S.; Mehrban, H.; Moharrery, A. Supplementation of different selenium sources during early lactation of native goats and their effects on nutrient digestibility, nitrogen and energy status. *J. Appl. Anim. Res.* **2018**, *46*, 64–68. [CrossRef]

75. Weiss, W.P. Selenium sources for dairy cattle. In Proceedings of the Tri–State Dairy Nutrition Conference, Fort Wayne, IN, USA, 2–3 May 2005.

76. Alimohamady, R.; Aliarabi, H.; Bahari, A.; Dezfoulian, A.H. Influence of Different Amounts and Sources of Selenium Supplementation on Performance, Some Blood Parameters, and Nutrient Digestibility in Lambs. *Biol. Trace Elem. Res.* **2013**, *154*, 45–54. [CrossRef]

77. Do, C.; Wasana, N.; Cho, K.; Choi, Y.; Choi, T.; Park, B.; Lee, D. The Effect of Age at First Calving and Calving Interval on Productive Life and Lifetime Profit in Korean Holsteins. *Asian Australas. J. Anim. Sci.* **2013**, *26*, 1511–1517. [CrossRef] [PubMed]

78. Hedaoo, M.; Khllare, K.; Meshram, M.; Sahatpure, S.; Patil, M. Study of some serum trace minerals in cyclic and noncyclic surti buffaloes. *Vet. World* **2008**, *1*, 71–72.

79. Kamada, H. Effects of selenium-rich yeast supplementation on the plasma progesterone levels of postpartum dairy cows. *Asian Australas. J. Anim. Sci.* **2017**, *30*, 347–354. [CrossRef] [PubMed]

80. Pursley, J.R.; Martins, J.P. Impact of circulating concentrations of progesterone and antral age of the ovulatory follicle on fertility of high-producing lactating dairy cows. *Reprod. Fertil. Dev.* **2011**, *24*, 267–271. [CrossRef] [PubMed]

81. Inskeep, E.K. Preovulatory, and postmaternal recognition effects of concentrations of progesterone on embryonic survival in the cow. *J. Anim. Sci.* **2004**, *82*, 24–39. [CrossRef]

82. Kamada, H.; Ikumo, H. Effect of selenium on cultured bovine luteal cells. *Anim. Reprod. Sci.* **1997**, *46*, 203–211. [CrossRef]

83. Bajaj, N.K.; Sharma, N. Endocrine Causes of Early Embryonic Death: An Overview. *Curr. Res. Dairy Sci.* **2011**, *3*, 1–24. [CrossRef]

84. Ganie, A.A.; Baghel, R.P.S.; Mudgal, V.; Aarif, O.; Sheikh, G.G. Effect of selenium supplementation on reproductive performance and hormonal profile in buffalo heifers. *Ind. J. Anim. Res.* **2014**, *48*, 27–30. [CrossRef]

85. Thatcher, W.; Santos, J.E.P.; Staples, C.R. Dietary manipulations to improve embryonic survival in cattle. *Theriogenology* **2011**, *76*, 1619–1631. [CrossRef]

86. Dimri, U.; Ranjan, R.; Sharma, M.C.; Varshney, V.P. Effect of vitamin E and selenium supplementation on oxidative stress indices and cortisol level in blood in water buffaloes during pregnancy and early postpartum period. *Trop. Anim. Health Prod.* **2010**, *42*, 405–410. [CrossRef]

87. Hosnedlova, B.; Kepinska, M.; Skalickova, S.; Fernandez, C.; Ruttkay-Nedecky, R.; Malevu, T.D.; Sochor, J.; Baron, M.; Melcova, M.; Zidkova, J.; et al. A Summary of New Findings on the Biological Effects of Selenium in Selected Animal Species—A Critical Review. *Int. J. Mol. Sci.* **2017**, *18*, 2209. [CrossRef] [PubMed]

88. Mézes, M.; Balogh, K. Prooxidant mechanisms of selenium toxicity—A review. *Acta Biol. Szeged.* **2009**, *53*, 15–18.

89. Hartikainen, H. Biogeochemistry of selenium and its impact on food chain quality and human health. *J. Trace Elem. Med. Biol.* **2005**, *18*, 309–318. [CrossRef] [PubMed]

90. Kabata-Pendias, A.; Pendias, H. *Biogeochemia Pierwiastk'ow'sladowych*; PWN Press: Warsaw, Poland, 1999.

91. Mihajlović, M. Toksicnost selena kod domaćih zivotinja [Selenium toxicity in domestic animals]. *Glas Srp Akad Nauka Med.* **1992**, *42*, 131–144.

92. Neve, J.; Favier, A. Selenium in medecine and biology. In Proceedings of the Second International Congress on Trace Elements in Medecine and Biology, Avoriaz, France, March 1988.

93. Claude, J.B. *Introduction à la Nutrition des Animaux Domestiques*; Tec & Doc/EM Inter: Paris, France, 2002.

94. Rogers, P.A.M.; Arora, S.P.; Fleming, G.A.; Crinion, R.A.P.; McLaughlin, J.G. Selenium toxicity in farm animals: Treatment and prevention. *Ir. Vet. J.* **1990**, *43*, 151–153.

95. James, L.F.; James, L. Shupe. Selenium Poisoning in Livestock. *Rangelands* **1984**, *6*, 64–67.

96. Nair, M.P.N.; Schwartz, S.A. Immunoregulation of natural and lymphokine-activated killer cells by selenium. *Immunopharmacology* **1990**, *19*, 177–183. [CrossRef]

97. Abuelo, A.; Alves-Nores, V.; Hernandez, J.; Muiño, R.; Benedito, J.L.; Castillo, C. Effect of parenteral antioxidant supplementation during the dry period on postpartum glucose tolerance in dairy cows. *J. Vet. Intern. Med.* **2016**, *30*, 892–898.

98. Rizzo, A.; Pantaleo, M.; Mutinati, M.; Minoia, G.; Trisolini, C.; Ceci, E.; Sciorsci, R.L. Blood and milk oxidative status after administration of different antioxidants during early postpartum in dairy cows. *Res. Vet. Sci.* **2013**, *95*, 1171–1174. [CrossRef]

99. Szpunar, J.; Lobinski, R.; Prange, A. Hyphenated techniques for elemental speciation in biological systems. *Appl. Spectrosc.* **2003**, *57*, 102–112. [CrossRef]

100. Najafnejad, B.; Aliarabi, H.; Tabatabaei, M.M.; Alimohamady, R. Effects of different forms of Selenium supplementation on production performance and nutrient digestibility in lactating dairy cattle. In Proceedings of the Second International Conference on Agriculture and Natural Resources, Kermanshah, Iran, 25–26 December 2013.

101. Kamada, H.; Hondate, K. Effect of dietary Selenium supplementation on plasma progesterone concentration in cows. *J. Vet. Med. Sci.* **1998**, *60*, 133–135. [CrossRef] [PubMed]

102. Hall, J.A.; Bobe, G.; Vorachek, W.R.; Kasper, K.; Traber, M.G.; Mosher, W.D.; Pirelli, G.J.; Gamroth, M. Effect of supranutritional organic selenium supplementation on postpartum blood micronutrients, antioxidants, metabolites, and inflammation biomarkers in selenium-replete dairy cows. *Biol. Trace Elem. Res.* **2014**, *161*, 272–287. [CrossRef] [PubMed]

Publisher's Note: MDPI stays neutral with regard to jurisdictional claims in published maps and institutional affiliations.

MDPI

St. Alban-Anlage 66

4052 Basel

Switzerland

Tel. +41 61 683 77 34

Fax +41 61 302 89 18

www.mdpi.com

Sustainability Editorial Office

E-mail: sustainability@mdpi.com

www.mdpi.com/journal/sustainability

www.ingramcontent.com/pod-product-compliance
Lightning Source LLC
LaVergne TN
LVHW070641170726
843515LV00004B/298